The McGraw-Hill Encyclopedia of Quality Terms & Concepts

James W. Cortada

John A. Woods

McGraw-Hill, Inc.

New York   San Francisco   Washington, D.C.   Auckland   Bogotá
Caracas   Lisbon   London   Madrid   Mexico City   Milan
Montreal   New Delhi   Paris   San Juan   Singapore
Sydney   Tokyo   Toronto

To Dora Cortada and Nancy Woods

Our partners and teammates

Library of Congress Cataloging-in-Publication Data
Cortada, James W.
    The McGraw-Hill encyclopedia of quality terms & concepts / James W. Cortada, John A. Woods.
        p.   cm.
    Includes index.
    ISBN 0-07-024099-X (hc)
        1. Total quality management—Encyclopedias.   2. Quality control—Encyclopedias.   I. Woods, John A.   II. Title.
HD62.15.C686   1995   95-15695
658.5′62′03—dc20   CIP

1 2 3 4 5 6 7 8 9 0 DOC/DOC 0 9 8 7 6 5

ISBN: 0-07-024099-X

The sponsoring editor for this book was James H. Bessent, Jr. It was produced under the direction of CWL Publishing Enterprises, Madison, Wisconsin. It was composed at Impressions, a division of Edwards Brothers, Inc.

Printed and bound by R.R. Donnelley and Company.

 This book is printed on recycled, acid-free paper
containing a minimum of 50% recycled de-inked fiber.

Preface

For many executives, supervisors, and employees, quality management can be a mixture of technicalities and platitudes. It seems weighed down in complicated statistics while promoting feel-good notions of cooperation and being nice to customers. There may be even be some truth in these impressions. However, there is also this fact: the best way to manage *is* through the practical implementation of quality principles and techniques. So we have prepared this encyclopedia.

What's our purpose? This reference has the purpose of taking the mystery out of the technicalities and putting some teeth in the platitudes of quality management. It gives you reliable and easy-to-understand definitions and examples of several hundred terms and concepts that make up the Total Quality Management lexicon. With this alphabetically organized guide on your desk or shelf, you have a convenient volume where you can look up and quickly read about all kinds of quality management concepts. We cover topics in such areas as statistical process control, activity-based costing, manufacturing, services, strategy planning, focus on customers, culture, measurements, and more.

Who are you? We envision our readers to be people like ourselves. Intelligent and involved (or at least interested) in making an organization operate better using quality management principles. You don't want to be overwhelmed by engineer-speak, but you want to at least appreciate the issues that technical people address in improving quality. You have a sense that enhancing the way people work together is important, but you would like to know more about how to make that happen. You know that meeting customer requirements is important, and you want to know more about that as well. In other words, you are on a TQM learning curve, and you'd like a reference you can turn to when you have a question about some issue. This book is that reference.

How does it work? If someone talks about an *operating characteristic curve* or *Six Sigma* or the *Taguchi loss function*, this is a place to learn quickly about such ideas. If someone talks about *customer delight, self-directed teams,* or *quality function deployment,* this is a place to learn about those ideas as well. Each of these concepts and hundreds more are covered here. When appropriate, we cross-reference other terms, to which you can turn for even further explanation.

Our definitions and discussions are not meant to be comprehensive. However, they are in enough detail to give you the basics. The goal is to provide just enough so you will feel comfortable with the idea. When appropriate, we include figures to illustrate concepts. Finally, most terms are followed by a short list of references where you can find still more information. Many of these references are McGraw-Hill titles, to which we had the fullest access, but we've sought to list the best books from all publishers.

To provide you with additional perspective, we have pulled quotes from many of the referenced works and put these in the margin next to our discussion. These quotes help further explain an idea. Or they give you some sense of one or another expert's feeling about the issue at hand. We think they add a lot of value to the book and hope you will, too.

Use it to leverage your time. While this book is meant as a reference, you may find it useful just to browse through. Like any encyclopedia, you can "page surf" through it just to see what's there. This exposes you to many new ideas that you may not have found out about otherwise. Looking up information on all the points included here in other, more specialized books is tedious and time consuming. Thus, this book is a good way to leverage your time to learn a little about a lot of TQM ideas.

Appendices

The references at the end of each entry tell you where to turn for more information. However, we have also included more detailed information in three appendices at the end of the book for furthering your study and involvement with this approach to management.

Appendix 1. This appendix includes a comprehensive annotated bibliography of most of the books we have used in compiling this encyclopedia. We have listed them by category. The annotations explain the value of the book from our perspective. We have noted those we especially recommend.

Appendix 2. This includes an annotated listing of magazines and journals that cover quality management issues. You'll find the name, address, subscription information, and a brief description of what distinguishes each item. We've grouped these by area of interest.

Appendix 3. This is an annotated listing of major organizations directly involved in promoting quality management tools, techniques, and culture and in certification activities. It is organized by geographical regions: The United States and Canada, Europe, and Asia and the Pacific. Many of these organizations offer training and other support activities for individuals and companies.

Acknowledgments

This book is a collaboration, and we want to publicly thank those who have helped make it a reality. Kevin Kelleher, a quality management consultant at Joiner Associates, read most of the manuscript and helped us make sure we didn't misrepresent any concepts. Don Ermer, the Procter & Gamble Professor of Quality Engineering at the University of Wisconsin, Madison, answered questions for us when we weren't sure of an issue. Our editor at McGraw-Hill, Jim Bessent, has been very supportive throughout our efforts.

We regularly took advantage of the library of references at the Madison Area Quality Improvement Network. We want to commend such organizations to you. They can provide you with much support in your quality journey.

The production of this book was handled by Impressions, a Madison book production house. We were fortunate to work with Mary Boss there, who very professionally guided this through the production process, with the help of Jane Tenenbaum who did the attractive design, and Pat Jones, who did the copyediting. Thanks to all of them.

Finally, we want to thank you for purchasing this book. We welcome any feedback on its value to you and suggestions for improvement of future editions. You can write to us at 3010 Irvington Way, Madison, WI 53713.

James W. Cortada
John A. Woods

Contents

ABC This is the acronym for activity-based costing, an approach to accounting. (See this term for more detail.)

Absolute Frequency This is another way of saying frequency. (See *Frequency*.)

Accelerated Life Test This refers to the testing of a product or component to determine its durability in normal use over its expected life.

> Accelerated life tests are of value only to the degree to which they correlate with actual-use life tests.
>
> *Armand V. Feigenbaum*

In undertaking accelerated life testing, the product is expected to fail in the lab just as it would in actual use. However this happens in much less time because of the harsh or rapid and repetitive stress to which the product is subjected. This helps companies make accurate predictions about the life of a component or product. For example, automobile companies use machines to open and close doors thousands of times to test the durability of hinges and locks.

For more information, see: Dev G. Raheja, *Assurance Technologies: Principles and Practices* (New York: McGraw-Hill, 1991).

Accept In manufacturing, this word means the decision that a quantity of materials, components, products, or services has met predetermined specifications for performance and quality and is therefore acceptable. Organizations often base acceptance on whether samples from a lot meet specific quality criteria.

1

For more information, see: Armand V. Feigenbaum, *Total Quality Control*, Third Edition, Revised (New York: McGraw-Hill, 1991); Joseph M. Juran and Frank M. Gryna, *Quality Planning and Analysis*, Third Edition (New York: McGraw-Hill, 1993).

Acceptable Component Failure Level This is the level of variance in the performance of a process or product that customers will find tolerable. Engineers normally arrive at this through testing or long-term observation of the process.

For more information, see: Eugene R. Carrubba and Ronald D. Gordon, *Product Assurance Technologies* (New York: McGraw-Hill, 1988); D.H. Stamatis, *Failure Mode and Effect Analysis: FMEA from Theory to Execution* (Milwaukee, WI: ASQC Quality Press, 1995).

Acceptable Error Rate This is the acceptable rate at which any process delivers acceptable compared to nonacceptable outputs. Reaching an acceptable error rate should never be considered a goal for any process. Rather, the goal should always be zero defects.

For further information, see: Charles A. Mills, *The Quality Audit* (New York: McGraw-Hill, 1989).

Acceptable Performance Level This is the lowest performance level that can be considered as acceptable for the process being audited.

The goal should always be to reach a specific acceptable performance level of not less than 100 percent. However, for auditing limited samples, it is a useful concept.

For more information, see: Charles A. Mills, *The Quality Audit* (New York: McGraw-Hill, 1989).

Acceptable Quality Level (AQL) This is the maximum percent of nonconforming pieces in a lot or group that, for the purposes of sampling inspection, can be considered satisfactory as the average for the output of a process. More broadly and less technically, it is the lowest quality level a customer will accept and is the standard a customer will use in judging a product or service.

Business people use this concept widely in both manufacturing and retail enterprises. Instead of inspecting all items, inspectors check a certain number of samples within lots or groups of items manufactured or delivered. If they meet specific quality performance standards, then we can assume the entire lot meets acceptable standards. Key to effective sampling is appreciation of the notion of variation in the quality characteristics of products and services delivered to customers by a process. (This insight is, perhaps, Deming's most important contribution to the application of statistical process control to improve management of quality). In TQM, AQL is becoming an outmoded concept because it creates a mindset that a small percentage of defects is acceptable. The continuous improvement approach to TQM suggests that management's goal should always be to eliminate all defects.

> For more information, see: H.G. Menon, *TQM in New Product Manufacturing* (New York: McGraw-Hill, 1992); Charles A. Mills, *The Quality Audit* (New York: McGraw-Hill, 1989); Eugene L. Grant and Richard S. Leavenworth, *Statistical Quality Control*, Sixth Edition (New York: McGraw-Hill, 1988); Armand V. Feigenbaum, *Total Quality Control*, Third Edition, Revised (New York: McGraw-Hill, 1991).

Acceptance Costs These are the costs involved with the testing and inspecting as well as the costs of administering an acceptance program. Increasing quality (that is, reducing defects by making processes more efficient and effective) reduces acceptance costs.

Acceptance Inspection This is the process of measuring, examining, testing, or otherwise comparing a product unit with specifications and requirements. Managers undertake this kind of inspection of components, products, or quality of a service to collect data to make decisions about whether to accept or release these components, products, or services based on predetermined accept/reject criteria.

Acceptable quality level plans favor the manufacturer since they provide high assurance of probable acceptance and do not take into account the customer's risk.

H.G. Menon

For example, the travel department in an organization might check on-time records, baggage lost, and consumer surveys to judge the acceptability of regularly using one airline. Manufacturers apply acceptance inspection to make decisions on whether to release batches of finished product.

For more information, see: Charles A. Mills, *The Quality Audit* (New York: McGraw-Hill, 1989); Armand V. Feigenbaum, *Total Quality Control*, Third Edition, Revised (New York: McGraw-Hill, 1991).

Acceptance Number (Attributes) In judging the acceptability of a lot or batch, this is the maximum number of nonconformities allowed within a sample, based on a count of the nonconformities. If the batch passes, that is, its amount of nonconformities falls at or below the acceptance number, this means the lot or batch meets requirements for quality or performance. If a batch does not pass, this usually indicates there is a problem with the process, which must be addressed.

For more information, see: Armand V. Feigenbaum, *Total Quality Control*, Third Edition, Revised (New York: McGraw-Hill, 1991).

Acceptance Procedure This refers to the predefined steps or procedures by which component or product ownership moves from supplier to customer.

Often well-documented and rigorous, an acceptance procedure will have a variety of inspection steps and documented accounts of quality. It may occur at various times. For example, acceptance of a component that goes into a product could be subject to an acceptance procedure four times: (1) as it comes out of a manufacturing location, (2) as it is accepted by another manufacturer that intends to use the component in another product, (3) when that larger product is manufactured, and (4) when a customer of the second product of which the component is a part also conducts an acceptance test.

For more information, see: Armand V. Feigenbaum, *Total Quality Control*, Third Edition, Revised (New York: McGraw-Hill, 1991).

Acceptance Sampling This is a technique for making decisions to accept or reject products or services based on samples. The term also refers to the procedures or processes by which managers make decisions to accept or reject the product or service by relying on inspection of samples. Acceptance sampling gives organizations a technique to avoid having to inspect, for example, all raw material or other supplies used in their own production processes or delivery of services.

Acceptance sampling is a mathematical process based on the fundamentals of statistical probability. This process assumes the random selection of samples will be representative of the entire group from which the samples come. It also assumes that all items of a group from which samples are drawn are the same within limits of variation or are homogeneous. For example, with a large lot of three-inch wood screws, this procedure assumes that by sampling, one can determine if all screws in that lot, within certain acceptable variation, are similar in length, appearance, and quality. Engineers can apply several mathematical formulas to determine the confidence level that sampling for inspection reflects the total quality of the lot.

> For more information, see: Armand V. Feigenbaum, *Total Quality Control*, Third Edition, Revised (New York: McGraw-Hill, 1991); Eugene L. Grant and Richard S. Leavenworth, *Statistical Quality Control*, Sixth Edition (New York: McGraw-Hill, 1988); Edward G. Schilling, *Acceptance Sampling in Quality Control* (Milwaukee, WI: ASQC Quality Press, 1982).

Acceptance Sampling Plan This is a plan for setting up the acceptable sampling technique that includes criteria setting sample size, and for accepting and rejecting a component, product, or service based on samples meeting prescribed quality criteria. Plans may involve single, double, sequential, chain, multiple, or skip-lot sampling techniques. With variable samples, management may deploy single, double, and sequential sampling techniques.

> The main advantage of sampling is economy. Despite some added costs for designing and administering the sample plan, the lower costs of inspecting only part of the lot result in an overall cost reduction.
>
> *Joseph M. Juran and Frank M. Gryna*

An acceptance sampling plan for the outputs of a process helps manufacturing personnel discover if and where quality problems exist in a process. Then, by employing the normal assortment of problem determination techniques, such as process mapping and root cause analysis, a manager can detect the source of a quality problem and then develop a method for eliminating it.

> For more information, see: Kiyoshi Suzaki, *The New Shop Floor Management: Empowering People for Continuous Improvement* (New York: The Free Press, 1993); Armand V. Feigenbaum, *Total Quality Control*, Third Edition, Revised (New York: McGraw-Hill, 1991); Joseph M. Juran and Frank M. Gryna, *Quality Planning and Analysis*, Third Edition (New York: McGraw-Hill, 1993).

Acceptance Sampling Scheme This refers to any set of procedures or steps that defines an acceptance sampling plan. There must be a demonstrable relationship between lot sizes, sample sizes, and acceptance criteria, such that this scheme gives results similar to 100 percent inspection. This term is most frequently used in manufacturing and processing environments.

> For more information, see: Armand V. Feigenbaum, *Total Quality Control*, Third Edition, Revised (New York: McGraw-Hill, 1991); Joseph M. Juran and Frank M. Gryna, *Quality Planning and Analysis*, Third Edition (New York: McGraw-Hill, 1993).

Acceptance Sampling System This refers to a collection of sampling schemes. Such a collection would also contain a description of the criteria used to select one scheme over another.

> For more information, see: Armand V. Feigenbaum, *Total Quality Control*, Third Edition, Revised (New York: McGraw-Hill, 1991); Joseph M. Juran and Frank M. Gryna, *Quality Planning and Analysis*, Third Edition (New York: McGraw-Hill, 1993).

Accept/Reject Criteria This refers to the criteria managers use to measure and decide whether or not the quality and performance of a product or service is acceptable. These criteria can be technical specifications (such as the level of size tolerances allowed in parts), contract terms and conditions (for example, completion of certain tasks for a specific amount of money), or performance of a process or service (for example, number of phone calls answered within one minute in a telephone center).

Implied in the term is the idea that all processes, products, and actions should have an assigned level of acceptable performance that one can measure. Consistent with the understanding that criteria grow out of process capabilities, technicians should periodically measure for acceptability to find out how well a process or item is meeting specifications.

For more information, see: H.G. Menon, *TQM in New Product Manufacturing* (New York: McGraw-Hill, 1992); Philip B. Crosby, *Quality Is Free* (New York: McGraw-Hill, 1979); Armand V. Feigenbaum, *Total Quality Control*, Third Edition, Revised (New York: McGraw-Hill, 1991).

Accreditation This is a formal process that checks and gives the accrediting organization's stamp of approval to the processes and operations of an organization or the skills of an individual. This is done, for example, by the Registrar Accreditation Board for ISO 9000 standards, the ASQC (American Society for Quality Control) for quality experts, and APICS (American Production and Inventory Control Society) for manufacturing specialists.

Accreditation can occur at the institutional or personnel levels, but it usually refers to institutions. For example, at the institutional level, ISO certification for individual manufacturing plants verifies that basic standards of quality processes are in place. At the individual level, certification can come following the completion of training and/or the accumulation of recognized experiences. The ASQC, for instance, conducts a series of classes in quality practices that, upon successful completion, leads to a quality certification. Individual certification in various disciplines (not just quality) is expanding worldwide. For example, it is rapidly becoming the norm in management consulting, and has been for decades for accountants (CPA), lawyers (passing the Bar exams), and medical doctors.

For more information, see: James L. Lamprecht, ISO 9000: *Preparing for Registration*, Milwaukee, WI: ASQC, 1992). Contact the American Society for Quality Control, Inc., 611 E. Wisconsin Ave., P.O. Box 3005, Milwaukee, WI 53201-3005, USA; telephone: (800) 248-1946 or (414) 272-8575 for information on individual accreditation.

Accuracy This refers to the closeness between the measures of an object or product performance and the actual true values of the physical conditions or values being measured.

Accuracy is important because just having a performance measure for a component or service does not guarantee a correct view of what is happening. In manufacturing, engineers have long understood this issue. In measuring processes, accuracy is likely to depend on the methodology for taking measurements and on then analyzing them (for example, what statistical tools managers employ).

It is useful to distinguish between the concepts of accuracy and precision. Accuracy is the closeness of the measurement taken to an actual or true value. Precision deals with the variation in the measuring device allowing for accuracy of measurement. A target shooting analogy, as shown in figure 1, demonstrates this difference. In the first target, the shots hit where they are aimed (accurate), but the aiming device (the sight) is not properly adjusted (imprecise). In the second target, the sight is properly adjusted (precise), but the aim is poor (inaccurate shooting technique). In the third target, both the aim and the sight are correct and the shots hit the bull's-eye (accurate and precise). In terms of

> Accuracy is often referred to as *the correctness of measurement.* Precision is considered to be the *fineness of measurement.* It incorporates reproducibility but not accuracy.
>
> *George L. Miller*
> *and LaRue L. Krumm*

FIGURE 1

The difference between accuracy and precision.

Accurate,
not precise

Precise,
not accurate

Accurate
and precise

measurement, this means that the measuring device is properly calibrated, and the measurement is properly taken.

> For more information, see: Joseph M. Juran and Frank M. Gryna, *Quality Planning and Analysis*, Third Edition (New York: Mc-Graw-Hill, 1993).

Action Limits Some people use this term in the same way as *control limits* established on any type of statistical quality control charts. When something occurs outside these control limits, technicians must take some action to figure out why and do something about it. Otherwise the process will become more and more unstable.

> For more information, see: W. Edwards Deming, *Out of the Crisis* (Cambridge, MA: MIT Center for Advanced Engineering Study, 1986); N. Logothetis, *Managing for Total Quality: From Deming to Taguchi and SPC* (Englewood Cliffs, NJ: Prentice Hall, 1992); Brian L. Joiner, *Fourth Generation Management* (New York: Mc-Graw-Hill, 1994).

Action Team (AT) This is a group of people authorized by management to carry out improvements to an existing process.

An example of an action team might be management assigning a half dozen individuals within an accounting department the task of improving or replacing an accounts receivable process. Management should usually select such teams based on their content knowledge, though there are exceptions. For example, an individual who lacks background in a field but understands process improvement techniques can be a valuable team member. Such individuals can bring a fresh approach to a project and bring up questions experienced people may forget to ask. For teams to be most effective, members should receive training in basic process improvement techniques and be authorized to make decisions regarding their project or problem.

> For more information, see: H. James Harrington, *Business Process Improvement* (New York: McGraw-Hill, 1991); Jack D. Orsburn et al., *Self-Directed Work Teams: The New American Challenge* (Burr Ridge, IL: Irwin Professional Publishing, 1990); Peter R. Scholtes, *The Team Handbook* (Madison, WI: Joiner Associates, 1988).

Activities In quality management, these are the steps (also tasks) in the execution of a process.

All processes are composed of interrelated tasks or steps. For example, consider the idea that all the activities associated with getting up in the morning to go to work are steps in a process. Then individual tasks or activities within that process could include (among several others) brushing your teeth, putting on your shoes, or strapping on your watch. Here is why activities are important: In process improvement or reengineering, it is critical to identify existing activities or steps to understand how a process currently operates and how well it works so managers can figure out how to improve it.

> For more information, see: H. James Harrington, *Business Process Improvement* (New York: McGraw-Hill, 1991); Sarv Singh Soin, *Total Quality Control Essentials* (New York: McGraw-Hill, 1992); James Champy and Michael Hammer, *Reengineering the Corporation* (New York: HarperBusiness, 1993); and many other books that deal with continuous improvement of processes.

Activity Analysis This is the analysis and evaluation of the performance of an activity or process. The intent is usually to quantify the expense of an activity and identify areas of cost-justified improvement.

Typical steps in activity analysis include (1) identifying nonessential activities (for example, excessive number of sign-offs) that have costs associated with them and that do not add value for customers, (2) analyzing significant activities (for example, the 20 percent of processes most important to the business or customers), (3) comparing best practice activities and costs, and (4) looking for links between activities to figure out how to better couple them and eliminate dead time between them.

> For more information, see: Peter B.B. Turney, *Common Cents: The ABC Performance Breakthrough* (Portland, OR: Cost Technology, 1991); Michael C. O'Guin, *The Complete Guide to Activity-Based Costing* (Englewood Cliffs, NJ: Prentice Hall, 1991).

Activity-Based Budgeting This is the process of preparing cost budgets using activity-based costing (ABC) methods to identify work loads and required resources. In this approach, managers assign costs to activities, rather than allocating costs by department. ABC assumes activities are what create costs in an enterprise. To make the organization more cost-effective, therefore, one looks at the relationship among activities, their costs, and the value delivered.

ABC strategies for accounting are new and most evident in companies inclined to manage processes rather than individuals. In such an environment, activity-based budgeting is a convenient way to identify the accounting data necessary to manage expenses. ABC has developed into a full body of accounting practices, yet this approach to budgeting and accounting is still in its infancy.

> For more information, see: Peter B.B. Turney, *Common Cents: The ABC Performance Breakthrough* (Portland, OR: Cost Technology, 1991); Michael C. O'Guin, *The Complete Guide to Activity-Based Costing* (Englewood Cliffs, NJ: Prentice Hall, 1991).

Activity-Based Costing (ABC) This is an accounting methodology by which organizations can more realistically measure the costs and performance of activities, processes, and cost objects. With this approach managers can assign costs to activities or processes based on their use of such resources as people and supplies. The same can be done for cost objects. One by-product of an ABC approach is the recognition that there is a causal relationship between cost drivers and activities.

Many experts believe ABC provides a better set of techniques for measuring costs. This is because it provides quantifiable answers to questions such as which activities and procedures are cost-effective to implement and add value to processes. It does this because it forces companies to track the actual costs of all activities involved in different business processes. This is different from conventional accounting practice, which assumes that products cause costs and does not

ABC reveals the problems you need to correct and the profitable opportunities that are available. Its ability to measure true business performance ensures that you improve where it counts the most—the bottom line.

Peter B.B. Turney

break out the costs of activities to better understand process efficiency. The ABC approach can be quite helpful in analyzing organizational processes for continuous improvement.

For more information, see: Peter B.B. Turney, *Common Cents: The ABC Performance Breakthrough* (Portland, OR: Cost Technology, 1991); Michael C. O'Guin, *The Complete Guide to Activity-Based Costing* (Englewood Cliffs, NJ: Prentice Hall, 1991).

Activity-Based Management (ABM) These are management practices devoted to the control and operation of activities or processes with the goal of improving operations. These practices focus on increasing the value received by customers and the profitability of the business through process management.

Undertaking ABM involves analyzing cost drivers, activities, and performance. Activity-based costing is the major source of both financial and performance data. This approach is becoming increasingly popular—although still in an early phase of acceptance—in those companies that focus significant attention to the management of processes.

For more information, see: Peter B.B. Turney, *Common Cents: The ABC Performance Breakthrough* (Portland, OR: Cost Technology, 1991); Michael C. O'Guin, *The Complete Guide to Activity-Based Costing* (Englewood Cliffs, NJ: Prentice Hall, 1991).

Activity Center In activity-based costing, this is an aggregation of activities related to a particular process or department. In any company, there are hundreds or even thousands of activities, and these would be difficult to track without some way to organize them. In book publishing, for example, a production department activity center would group information about all activities involving copyediting, design, typesetting, art creation, and printing.

The activity center approach allows for a meaningful way not only to group related activities but also to better understand the costs and value delivered by the activities in this center.

For more information, see: Peter B.B. Turney, *Common Cents: The ABC Performance Breakthrough* (Portland, OR: Cost Technology, 1991).

Activity Cost Pool This term refers to all the costs that can be traced to a single activity. The activity cost pool includes all the costs associated with a particular activity. Tracing all costs in this way helps identify (1) how costs escalate when an activity is poorly executed and (2) how processes may be improved.

For example, if, in producing a book, an editor doesn't do a good job of copyediting a manuscript, the activity pool for this would also include the correction of mistakes in typesetting traceable to this activity.

For more information, see: Peter B.B. Turney, *Common Cents: The ABC Performance Breakthrough* (Portland, OR: Cost Technology, 1991).

Activity Driver This is always a factor employed in assigning the cost from an activity to a cost object, using ABC techniques. It is a measure of how frequently and intensely a cost object employs an activity.

For example, if nailing siding onto a house is the activity performed, then one would consider the number of sidings nailed to be an activity driver. Another example would be the re-designing of a product, which would yield activity driver "engineering changes." Though they vary in importance and cost, there are three levels of activity drivers: unit, batch, and product. ABC always assigns the cost of unit activities by measuring product units. Thus, for instance, the number of direct labor hours required for a task (for example, screwing two components together) is a measure of an operator's effort. In ABC, activity drivers are found in direct labor hours, machine hours, cost of materials, and in product units.

For more information, see: Peter B.B. Turney, *Common Cents: The ABC Performance Breakthrough* (Portland, OR: Cost Technology, 1991).

Activity Symbol This is a rectangle in a flow diagram and denotes an activity.

Adequacy of Standards This is a manufacturing term that refers to the ability of a standard to be used for calibrating the accuracy of a gage or instrument used to measure quality or performance. It is best if such standards are traceable to the National Bureau of Standards (NBS). The term also appears when dealing with the measurement of processes and procedures.

> For more information, see: James Robert Taylor, *Quality Control Systems* (New York: McGraw-Hill, 1989).

Administrative Time In quality control terms, this is the time during which a process or product fails to perform to specifications. Everyone then waits for management to take corrective actions so they can get back to acceptable levels of performance. Administrative time may also refer to non-value added activities that generate costs but not value.

Understanding how much administrative time exists, what causes it, and what goes on during this period is an important aspect of continuous improvement. Obviously you want to reduce this downtime. A common improvement strategy (especially for reducing "cycle time") has long been to reduce administrative time, especially the amount of time spent waiting for administrative action. This dead time makes cycle times longer. Measuring administrative time is a technique for understanding where and how quality failures occur. It also is a useful source of information for identifying the costs of poor quality. This is because managers are often involved in activities concerning the correction of mistakes.

> For more information, see: Gerard H. Gaynor, *Exploiting Cycle Time in Technology Management* (New York: McGraw-Hill, 1993); Joseph M. Juran and Frank M. Gryna, *Quality Planning and Analysis*, Third Edition (New York: McGraw-Hill, 1993).

Affinity Diagraming This is a technique for organizing a variety of subjective data (such as opinions) into categories based

> Recent experience makes it clear that quality concepts apply equally to administrative and support activities, all of which have customers—some internal, some external.
>
> *Joseph M. Juran and Frank M. Gryna*

on the intuitive relationships among individual pieces of information. It is often used by groups to find commonalities among concerns and ideas on any subject from members.

In this technique, you organize information into logical groups. For example, in a brainstorming meeting, the leader may ask people to identify the causes of world hunger. They then typically put each idea on a "Post-it" note and stick them all on a wall. The next step is to cluster the notes for each category of cause. That would lead you to groupings of notes that center on causes of hunger such as bad weather, poor national agricultural policies, and so forth. The value of an affinity diagram is that it leads you to several basic, critical ideas (that suggest actions you might take). Figure 2 illustrates an affinity diagram for world hunger.

FIGURE 2
Affinity diagram.

AFFINITY DIAGRAM FOR CAUSES OF WORLD HUNGER

Note: Categories of ideas are numbered and related ideas are then listed within each category.

1. **OVERPOPULATION**
   Poor birth control
   No means for supporting large families
   Values promoting large families in conflict with economic realities

2. **BAD WEATHER**
   Droughts
   Flood

3. **POLITICAL PROBLEMS**
   Wars
   Political elites exploit the poor
   Poorly developed economic infrastructure
   Poorly developed distribution systems

4. **AGRICULTURAL POLICIES AND METHODS**
   No use of fertilizer
   No crop rotation
   Little cooperation
   Low productivity farming methods
   Soil erosion

For more information, see: N. Logothetis, *Managing for Total Quality: From Deming to Taguchi and SPC* (Englewood Cliffs, NJ: Prentice Hall, 1992); PQ Systems, Inc., *Total Quality Transformation Improvement Tools* (Miamisburg, OH: Productivity Quality Systems, Inc., 1994, 800-777-3020); Nancy R. Tague, *The Quality Toolbox* (Milwaukee, WI: ASQC Quality Press, 1995).

Agenda This is a plan for conducting a meeting. A well-planned and executed meeting process is integral to the operations of any organization, and TQM recognizes this. A complete agenda will include (1) the topic of discussion, (2) the person in charge of the discussion, (3) the amount of time devoted to the discussion, and (4) the technique employed in the discussion.

An effective meeting process includes (1) designating someone to be in charge of the meeting, (2) another person to take

FIGURE 3

An example of an effective meeting agenda.

AGENDA

Purpose:	Update on Marketing Plan
Location:	Third Floor Conference Room
Date:	July 18, 19--
Time:	2:30-4:30 P.M.
Participants:	Jack Allison, Pat Rochester, Ginger Brown, Steve Peterson, Roger Murphy
Items to Bring:	September marketing plan update document

TOPIC		TIME	LEADER	METHOD	OUTCOME
1.	Check-In	5 minutes	Everyone	Round Robin	
2.	Review of Plan	15 minutes	Pat	Discussion	
3.	June Results	15 minutes	Jack	Presentation	
4.	New Advertising Campaign	25 minutes	Ginger	Brainstorm Discussion	Plan
5.	Revise Marketing Plan	50 minutes	Jack	Discussion	Decision
6.	Forward Actions	10 minutes	Jack	Discussion	Decision
7.	Evaluation of Meeting	10 minutes	Jack	Round Robin	

notes and distribute these promptly, (3) adhering to an agenda, and (4) applying good meeting practices. In the quality world, it is also common to see a meeting end with a brief discussion of how the meeting went and how to make future ones more productive. Figure 3 illustrates an effective agenda.

> For more information, see: Peter R. Scholtes, *The Team Handbook* (Madison, WI: Joiner Associates, 1988); *Running Effective Meetings* (Madison, WI: Joiner Associates, 1993); Michael Doyle and David Straus, *How to Make Meetings Work,* (New York: Jove Books, 1976).

Agile Manufacturing This is a set of flexible manufacturing processes that leads to the rapid design and manufacture of products and provides customers with what they want, when and where they want it. Using agile manufacturing rather than traditional manufacturing methods, firms can offer more customized products at competitive prices with no compromise in quality or damage to the environment.

It was originally conceived in the Japanese auto industry seeking to develop the capability to manufacture automobiles for individual customers in three days. It includes tactics and processes characterized by an ability to design and manufacture multiple products quickly.

> For more information, see: James W. Cortada and John A. Woods, *Quality Yearbook 1994* (New York: McGraw-Hill, 1994); B. Joseph Pine II, *Mass Customization: The New Frontier in Business Competition* (Boston: Harvard Business School Press, 1993).

Agility Another term for agile manufacturing, it means delivering to customers what they want, when and where they want it, at a reasonable price. It also means doing this with no compromises on quality or damage to the environment. Mass customization is another term for agility.

Agile manufacturing has led to more aggressive design and production schedules with reduced cycle time and more varied product design.

> For more information, see: R.W. Hall and J. Nakane, *Manufacturing Battlefield of the 90s* (Wheeling, IL: Association for Manufacturing

At a ridiculous extreme, agility means meeting any need for change instantly. It's a 21st century idea for manufacturing excellence— if such a milieu can still be called "manufacturing."

Robert W. Hall

Excellence, 1990); B. Joseph Pine II, *Mass Customization: The New Frontier in Business Competition* (Boston: Harvard Business School Press, 1993); James W. Cortada and John A. Woods, *The Quality Yearbook 1994* (New York: McGraw-Hill, 1994).

American Quality Foundation (AQF) This is an independent organization established under the American Society for Quality Control to foster the application of quality practices by business and public officials. It also has a research and development mission, responsible for programs to promote U.S. business competitiveness.

It is best known for a series of publications in recent years surveying quality practices in Europe, Asia, and the United States. Its seminal publication (published jointly with Ernst & Young) is the *International Quality Study* (Cleveland, OH: Ernst & Young, 1991). To contact AQF, write to the American Quality Foundation, 787 Seventh Ave., New York, NY 10019, USA, telephone: (212) 773-5600.

American Society for Quality Control (ASQC) The world's leading nonprofit professional association dedicated to the promotion and application of quality-related practices in both public and private sectors. It has over 96,000 individual members and over 700 corporate members in 64 countries.

It sponsors many seminars covering all aspects of quality in one to three-day formats. It also publishes *Quality Progress* (a monthly magazine on quality applications that goes out to over 135,000 subscribers), several other periodicals, and books on quality (ASQC Quality Press). Address: 611 E. Wisconsin Ave., P.O. Box 3005, Milwaukee, WI 53201-3005, USA, telephone: (800) 248-1946 or (414) 272-8575.

For more information, see: *Quality Progress* or write ASQC for a membership packet.

Analysis of Means (ANOM) This is a statistical technique used when running experiments to identify problems and/or capabilities of an industrial process to deliver an end product with the desirable characteristics. It helps an operator identify the best combination of factors in a process for generating the desired outcome, such as tensile strength of a metal component or similar output characteristics. In other words, it is a way to determine by experiments and statistical calculations which process methods and combination of materials (the means) are most likely to deliver the best output (end).

For more information, see: H.G. Menon, *TQM in New Product Manufacturing* (New York: McGraw-Hill, 1992); Wayne A. Taylor, *Optimization and Variation Reduction in Quality* (New York: McGraw-Hill, 1991); Ellis R. Ott and Edward G. Schilling, *Process Quality Control*, Second Edition (New York: McGraw-Hill, 1990).

Analysis of Variance (ANOVA) This is a method for breaking down and analyzing the total variation in the outputs of any process, understanding the causes of this variation, and then assessing their significance. In designing a process, engineers devise experiments and then use ANOVA techniques to identify and analyze different causes of variation. The goal is to come up with a process in which variation in outputs will be minimized. This happens when variation in the processes themselves is also minimized. ANOVA techniques include involved statistical analysis. (See also *Poka-Yoke.*)

For more information, see: N. Logothetis, *Managing for Total Quality: From Deming to Taguchi and SPC* (Englewood Cliffs: Prentice Hall, 1992); H.G. Menon, *TQM in New Product Manufacturing,* (New York: McGraw-Hill, 1992); Ellis R. Ott and Edward G. Schilling, *Process Quality Control* (New York: McGraw-Hill, 1990).

Analysis Sample This refers to a randomly selected group of components from a complete group, batch, or lot of components, products, or services. Using statistical techniques to test samples representing the entire group of items is a reliable method for understanding the quality of all the items in that group.

For more information, see: George L. Miller and LaRue L. Krumm, *The Whats, Whys & Hows of Quality* (Milwaukee, WI: ASQC Quality Press, 1992).

Anatomy of Processes This is a biological metaphor for describing the linkage of many operations, such as steps, procedures, or processes that collectively result in the manufacture of a product or delivery of a service.

Annual Internal Training Expenditure This phrase refers to all expenditures for training, including salaries, office space, materials, information technology, and so forth. Typically these do not include the salaries of students during their training. Training expenditures can be a measurement of the relative commitment to improve employee skills in areas such as process management and statistical process control. Quality award programs frequently use such data to indicate corporate commitment to quality programs. Many companies benchmark this area.

> For more information, see: Stephen George, *The Baldrige Quality System* (New York: John Wiley & Sons, 1992); Thomas H. Berry, *Managing the Total Quality Transformation* (New York: McGraw-Hill, 1991).

AOQ This is an acronym for average outgoing quality. (See full term for definition.)

AOQL This is an acronym for average outgoing quality limit. (See full term for definition.)

Application The term has two meanings. The traditional one is from information systems. It describes a software program that performs a service (for example, order entry or word processing). The term also stands for the deployment of training knowledge in daily work. For example, if a worker learned how to apply statistics to a process, the use of those statistical techniques would be considered an application of statistical process control.

Appraisal Costs These include all costs for inspecting, testing, and quality audits to ensure that the outputs of a process meet customer specifications and performance requirements. In other words, these are the costs of checking at the end of a process to make sure outputs meet specifications.

Appraisal costs along with the costs of internal failure, external failure, and prevention represent four useful components for understanding the cost of quality. The importance of computing quality costs is that they help a company identify where it may improve, and they provide a way to measure progress in making improvements. The overall goal is to continually refine the efficiency and effectiveness of processes so as to reduce these costs, especially those involved with internal and external failure. TQM provides a variety of techniques that emphasize prevention of the causes of defects. (See also *Prevention.*)

> For more information, see: The Ernst & Young Quality Improvement Group, *Total Quality: An Executive's Guide for the 1990s* (Burr Ridge, IL: Irwin Professional Publishing, 1990); H. James Harrington, *The Improvement Process* (New York: McGraw-Hill, 1987); H. James Harrington, *Total Improvement Management* (New York: McGraw-Hill, 1994).

Approach This is one of three dimensions in Baldrige scoring used to evaluate performance of a company. (The other two are deployment and results.) This evaluative perspective refers to all the methods and techniques used by a company to achieve the objectives of the Baldrige Criteria.

Characteristics of the approach include describing to what extent a method, for example, is systematic, integrated, applied consistently, prevention-based, and the extent deployed throughout the organization.

> For more information, see: Mark Graham Brown, *Baldrige Award Winning Quality* (White Plains, N.Y.: Quality Resources and Milwaukee, WI: ASQC Quality Press, 1991, published annually); Christopher W.L. Hart and Christopher E. Bogan, *The Baldrige* (New York: McGraw-Hill, 1992); Stephen George, *The Baldrige Quality System* (New York: John Wiley & Sons, 1992).

Approved Component In manufacturing, this term describes parts used in the manufacture of a product that have successfully passed a quality test and have been deemed suitable for use.

Appraisal costs are all the costs expended to determine whether an activity was done right every time.

H. James Harrington

Approved Supplier This is a supplier who has passed a quality approval process. An approved supplier has quality control procedures acceptable to a customer (for example, is ISO 9000 certified), and therefore has the designation of supplier of choice in acquiring components, products, or services.

As organizations move toward using fewer and fewer suppliers, the requirement for suppliers to deliver according to negotiated standards becomes critical. This use of approved suppliers is now a common tactic in most industries. This approach makes suppliers part of the extended value chain of activities of a company. They become responsible for delivering components on time, for anticipating demand for more product, and for participating in the improvement of their customer's processes. Many companies now have certification procedures for suppliers and even awards for those suppliers that prove to be the most reliable.

> For further information, see: Richard J. Schonberger, *Building a Chain of Customers* (New York: The Free Press, 1990); James F. Cali, *TQM for Purchasing Management* (New York: McGraw-Hill, 1993).

Approved Supplier Program (ASP) This is always a formal program or process managed by customers by which they systematically approve vendors to be suppliers of parts, products, and services.

Such programs include statements of responsibility, levels of quality required in products and services delivered, and measurement schemes to ensure conformance. For both parties, there are many advantages to such a program. For customers, it lowers the costs of dealing with multiple vendors and reduces the expense of quality inspection of components and services. For suppliers, this virtually guarantees higher volumes of business as long as they meet quality standards. ASPs usually include steps to encourage dialogue between supplier and customer on how to continuously improve the operations of each.

For more information, see: James F. Cali, *TQM for Purchasing Management* (New York: McGraw-Hill, 1993).

Assembly Tree This is a process form that depicts inputs from many suppliers converging into subassemblies and assemblies. The illustration appears like a tree.

Often used within manufacturing environments, it is particularly common in automotive environments and is very evident in the manufacture of household appliances and a broad range of electronics. Leaves and roots represent suppliers. Suppliers can be vendors or departments within the company.

Assessment This refers to those steps taken to accept or reject parts, products, or services based on their fitness for use. The term also suggests any analysis of the performance of a process or quality of a product. It can refer to the activities involved in identifying the practices, attitudes, culture, and activities that either enhance or hinder the achievement of continuous improvement of quality in any organization.

For more information, see: Joseph M. Juran and Frank M. Gryna, *Quality Planning and Analysis*, Third Edition (New York: McGraw-Hill, 1993); Eugene H. Melan, *Process Management* (New York: McGraw-Hill, 1993); The Ernst & Young Quality Improvement Group, *Total Quality: An Executive's Guide for the 1990s* (Burr Ridge, IL: Irwin Professional Publishing, 1990).

Assignable Cause This is any cause of variation in a process or any performance that an employee can define. It may refer to both special causes and common causes, though it is usually associated with special causes.

If data can point to a cause, it is assignable. Consider, for example, a telephone system. If the data shows that when a phone rings more than seven times the caller hangs up, you have an assignable cause—the seven rings.

For more information, see: Ellis R. Ott and Edward G. Schilling, *Process Quality Control*, Second Edition (New York: McGraw-Hill, 1990).

Associate This is a common term used to refer to employees in an organization that has implemented TQM. It recognizes a mutual respect for the roles of all people in a firm and the importance of cooperation and teamwork in the execution of organization processes. It often suggests the idea of empowerment and the institution of self-managed teams.

> What makes the term "associate" significant is that our people are treated with respect and care, hour after hour, day in, day out. A person's attitude about his company is a direct reflection of the experiences he has within that organization.
>
> *Hal F. Rosenbluth*

For more information, see: Hal F. Rosenbluth and Diane McFerrin Peters, *The Customer Comes Second* (New York: William Morrow, 1992); Jack D. Orsburn et al., *Self-Directed Work Teams: The New American Challenge* (Burr Ridge, IL: Irwin Professional Publishing, 1990); William C. Byham and Jeff Cox, *ZAPP: The Lightning of Empowerment* (New York: Fawcett Columbine, 1988).

Attribute This is a piece of data that is countable in whole numbers. An attribute is always seen as either conforming or not conforming to some standard of quality or performance. Attribute control charts are developed by taking counts of nonconforming characteristics of process outputs. These control charts help show the capabilities of a process.

For more information, see: George L. Miller and LaRue L. Krumm, *The Whats, Whys & Hows of Quality Improvement* (Milwaukee, WI: ASQC Quality Press, 1992); Hy Pitt, *SPC for the Rest of Us* (Reading, MA: Addison-Wesley, 1994).

Attribute Data This is quantitative data obtained by counting the number of defects or flaws in the outputs of a process. It also includes qualitative data on one of two conditions, such as acceptable/not acceptable, go/no go, conforming/nonconforming, pass/fail, or present/absent.

Attribute data can be recorded on control charts using measurements as percents, number of affected units, counts, counts-per-unit, quality scores, and demerits. The purpose of such data is to help individuals understand process capabilities and make decisions for improvements. For example, FedEx may collect attribute data on what percent of overnight packages are delivered by 10:00 A.M. the next day. The attribute being counted is either yes, it was delivered on time or no, it was not. An appliance manufacturer may take counts of the number of scratches on all refrigerators as they come off the assembly line or the total number of refrigerators coming off the line with some kind of flaw or defect.

> For more information, see: Ellis R. Ott and Edward G. Schilling, *Process Quality Control*, Second Edition (New York: McGraw-Hill, 1990); Henry L. Lefevre, *Quality Service Pays: Six Keys to Success!* (Milwaukee, WI: ASQC Quality Press, 1989); W. Edwards Deming, *The New Economics: For Industry, Government, Education* (Cambridge, MA: MIT Center for Advanced Engineering Study, 1993); John L. Hradesky, *Productivity and Quality Improvement* (New York: McGraw-Hill, 1988).

Attributes, Method of This is a method selected for determining what defects or flaws one will count and how the count will be undertaken. It does not measure how far a parameter varies from some nominal value, just whether it is present or not and how much.

> For more information, see: Ellis R. Ott and Edward G. Schilling, *Process Quality Control*, Second Edition (New York: McGraw-Hill, 1990); John L. Hradesky, *Productivity and Quality Improvement* (New York: McGraw-Hill, 1988).

Audit, Quality This includes all the actions taken by an organization to assure that its processes are delivering quality outputs and to determine how well these processes function. As with accounting audits, the quality audit anticipates that procedures are in place and that the organization has documented these procedures.

Quality audits are an integral part of certification programs, such as ISO 9000, and quality award programs, such as the Baldrige and the Deming Awards. Organizations can undertake audits on the quality of components or products, performance of a process, or degree of adherence to customer requirements. Increasingly organizations also conduct audits of quality improvement processes, plans, and strategies for whole organizations. Audits are considered critically important processes in organizations dedicated to continuous quality improvement.

A properly conducted quality audit is a positive and constructive process. It helps prevent problems in the organization being audited through the identification of activities liable to create future problems.

Charles A. Mills

There has been considerable debate about when to conduct audits. Historically organizations have executed audits after they make a product or render a service, following the tradition established by accounting audits. As the understanding of

FIGURE 4

An example of a checklist for undertaking a quality audit of a process. Adapted from John L. Hradesky, *Productivity and Quality Management* (New York: McGraw-Hill, 1988).

Process Audit Checklist

Plant: Date:
Product or Area: Auditor:
Station: Operator:

Element No. Element Description	Verification			Diagnostic			Comments
	A	M	U	A	M	U	
1. Are documents complete?							
2. Are documents clear?							
3. Are documents correct?							
4. Are operations in proper sequence?							
5. Are reword and reinspect instuctions included?							
6. Are interactions included for defects to be reinspected?							
7. What is the latest revision?							
8. Are reference documents included where applicable?							
9. Are necessary support documents available? (For example, variations, workmanship standards, photos, sketches, minimum acceptable samples.)							
10. Does the manufacturing instruction adequately address all quality issues called for on the inspection instruction?							
11. Does the inspection instruction adequately represent the product specification and workmanship standard?							

*A = Acceptable, M = Marginal, U = Unacceptable

process management has expanded, organizations are now auditing their processes for efficiency and effectiveness.

Many experts advocate undertaking audits in the early stages of a process. This is because early audits call attention to possible problems that downstream might lead to quality defects they could avoid by fixing something upstream. Figure 4 illustrates a typical audit checksheet for a manufacturing process.

> For more information, see: Charles A. Mills, *The Quality Audit* (New York: McGraw-Hill, 1989); Joseph M. Juran and Frank M. Gryna, *Quality Planning and Analysis,* Third Edition (New York: McGraw-Hill, 1993); John L. Hradesky, *Productivity and Quality Improvement* (New York: McGraw-Hill, 1988).

Augmented Product　This is a way of defining the total bundle of benefits customers receive when they purchase a product. Augmented products include such benefits as extended warranties, trusted brand names, prestige image, service after the sale, and various benefits that extend beyond the tangible product itself. All of these have value to customers.

Exceeding customer expectations is a concept that plays an important part in many marketing and product strategies. Using the augmented product concept, you set a high level of expectation and then seek to consistently exceed that committed level of performance. For example, consider a furniture store that guarantees delivery of on-order chairs in six weeks when the industry standard is eight weeks. To exceed customer expectations, this store's managers may have developed a process to make deliveries within five weeks. This faster delivery is part of the added value that makes one furniture store more competitive than another.

To make such a strategy work in the long run requires two well-executed activities: (1) an accurate understanding of what customers value, and (2) an ability to anticipate and deliver new levels of customer delight. As you exceed expectations, your customers will then come to expect even higher levels of performance. Thus, for example, if a supplier routinely delivers

> The *new competition* is not between what companies produce in their factories, but between what they add to their factory in the form of packaging, services, advertising, customer advice, financing, delivery arrangements, warehousing, and other things that people value.
>
> *Theodore Levitt*

furniture in five weeks, its regular customers will come to expect that delivery time as normal. To exceed these expectations and maintain a competitive edge, you would have to deliver even faster or perform some other service. In other words, the necessity for continuously improving the whole package of benefits is always there.

> For more information, see: Valarie A. Zeithaml, A. Parasuraman, and Leonard L. Berry, *Delivering Quality Service: Balancing Customer Perceptions and Expectations* (New York: The Free Press, 1990); Laura A. Liswood, *Serving Them Right: Innovative and Powerful Customer Retention Strategies* (New York: Harper & Row, 1990); Carl Sewell, *Customers for Life* (New York: Doubleday, 1990); Karl Albrecht, *The Only Thing That Matters: Bringing the Power of the Customer Into the Center of Your Business* (New York: HarperBusiness, 1992); Eberhard E. Scheuing and William F. Christopher, *The Service Quality Handbook* (New York: Amacom, 1993).

Autonomous Department This is the department depicted on any production chart that acquires various inputs as components and converts them into finished goods or services. If one department performs all these manufacturing and delivery activities, that functional group is an autonomous department.

Autopsy In business this is a metaphor for any analysis carried out on the processes that deliver products or services to identify sources of specific deficiencies or problems.

Availability This means a machine, process, or person is ready and able to perform a function under a set of predetermined conditions. It is the ratio of uptime divided by the total of uptime and downtime. Uptime is when something is in active use (such as a machine), while downtime is when it is not able to function because it is being repaired or waiting to be repaired. A goal of quality management is to increase uptime and reduce downtime. This happens by focusing on preventive maintenance rather than repair after a breakdown.

Managers of computer centers, for example, track the amount of time a computer is performing (is "up") and continuously look for ways to get the uptime higher. This is particularly important if many people are relying on a machine or process (for example, the central computer). *Process availability* is a more recent measure of performance. This means all steps in a process are functioning so as to eliminate bottlenecks at any one or more steps.

For more information, see: H.G. Menon, *TQM in New Product Manufacturing* (New York: McGraw-Hill, 1992).

Average Chart This refers to a control chart on which we note the subgroup averages of a measured variable to evaluate a process's stability. Subgroups are samples of usually around five items from a group of process outputs. An average chart is also called an "X-bar" chart, which is shown symbolically as an X with a line or bar over it.

> The power of this type of chart is that as soon as the out of control condition is detected, the worker can take action to correct the situation. This prevents additional defective output from being produced.
>
> *William Lareau*

In using this chart, you calculate the average of some measured variable (such as the width of a sample of rolled steel or the gap in a sample of spark plugs) of a subgroup and record it on a control chart. You continue to take similar subgroups of averages and record each of these on the chart as well. You then calculate the mean of these averages, and this is the center line of the chart. Using this information, you can calculate the *upper* and *lower control limits* (see these entries) to determine both process capability and whether the process is in or out of control.

Average or X-bar charts are always employed with range charts. Range charts record the difference between the highest and lowest measure of the subgroup sample. (For example, if in a subgroup of five items, the highest measure is .14 and the smallest is .08, the range is .06 (.14−.08). You then calculate a mean of the ranges recorded and the upper and lower control limits. The average and range charts work together to help you assess how a process is operating. Figure 5 shows a standard form for developing an average and range chart. (See also *Range Chart.*)

FIGURE 5
Form for $\bar{X}$-R chart.

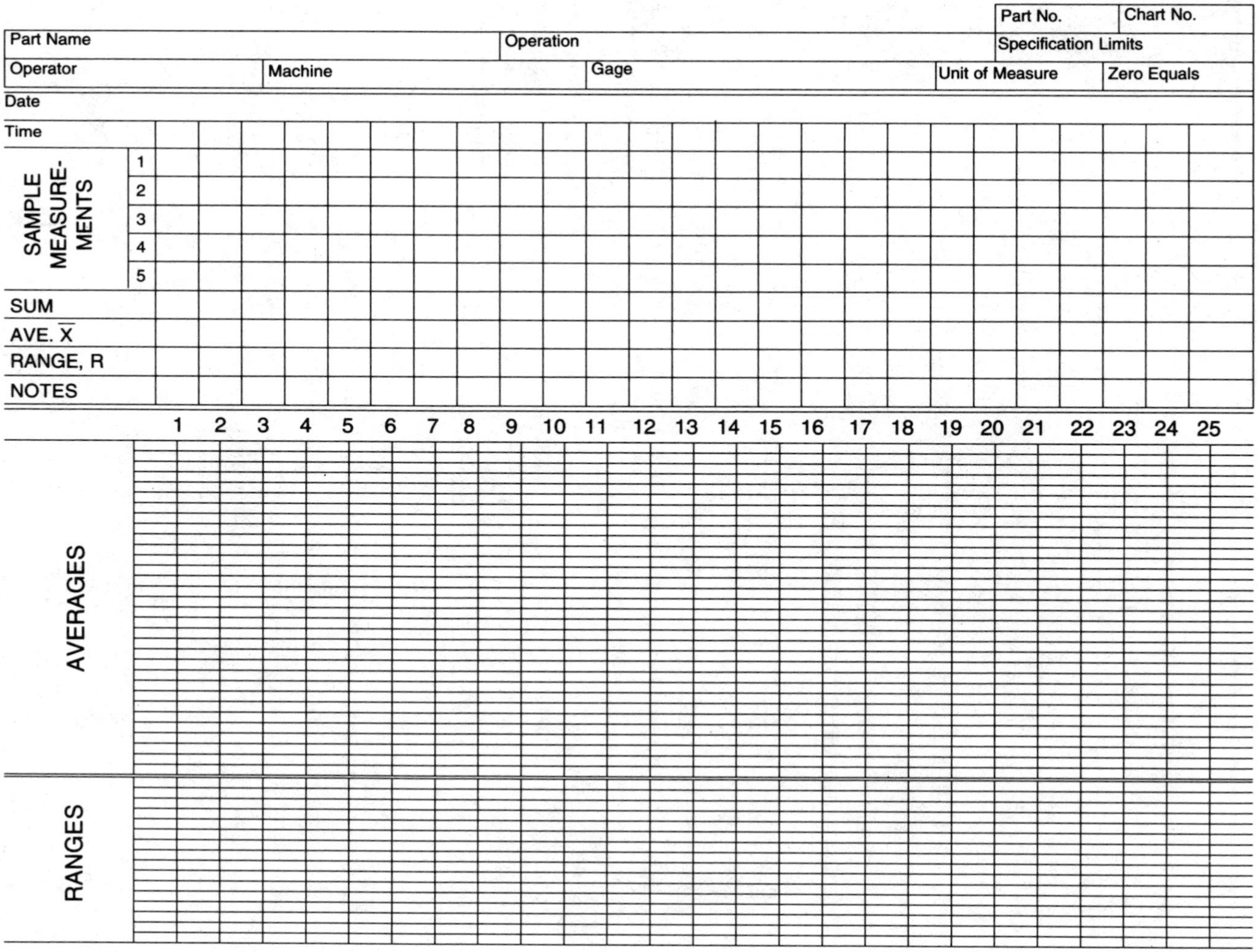

Part Name		Operation			Part No.	Chart No.
Operator	Machine		Gage		Specification Limits	
					Unit of Measure	Zero Equals

Date

Time

| SAMPLE MEASURE-MENTS | 1 | 2 | 3 | 4 | 5 |

| SUM |
| AVE. $\bar{X}$ |
| RANGE, R |
| NOTES |

1 2 3 4 5 6 7 8 9 10 11 12 13 14 15 16 17 18 19 20 21 22 23 24 25

AVERAGES

RANGES

For more information, see: Henry L. Lefevre, *Quality Service Pays: Six Keys to Success!* (Milwaukee, WI: ASQC Quality Press, 1989); Hy Pitt, *SPC for the Rest of Us* (Reading, MA: Addison-Wesley, 1994); John L. Hradesky, *Productivity and Quality Improvement* (New York: McGraw-Hill, 1988); and many other books that include a discussion of quality management statistical tools.

Average Outgoing Quality (AOQ) This refers to the maximum percent of defective products that can go to a customer after inspection performed by an acceptance sampling plan. AOQ helps reassure customers that suppliers will deliver products that will not include more defectives than expected.

AOQ requires 100 percent inspection if defectives in a batch or lot of products exceed some minimum standard. When this takes place, the supplier then replaces all defectives with acceptable units. (See *Average Outgoing Quality Limit* for more on this).

The Average Outgoing Quality method is required, for example, by U.S. military purchasers of their suppliers. Companies supplying goods to manufacturers may also employ this approach to ensure an acceptable (pre-agreed to) level of quality of parts, raw materials, or finished products.

For more information, see: Henry L. Lefevre, *Quality Service Pays: Six Keys to Success!* (Milwaukee, WI: ASQC Quality Press, 1989), H.G. Menon, *TQM in New Product Manufacturing* (New York: McGraw-Hill, 1992); Armand V. Feigenbaum, *Total Quality Control*, Third Edition, Revised (New York: McGraw-Hill, 1991).

Average Outgoing Quality Limit (AOQL) This refers to the average maximum percent of defectives that may go to a customer after inspection by some chosen sampling plan.

U.S. military purchasers led the way in the use of this approach to negotiate levels of quality in products they acquire. Many purchasers in large manufacturing operations also use AOQL because it tells them what could be the worst average quality they can expect from suppliers.

It works this way: Customers specify the average percent of defectives they will accept in a batch or lot. If the defectives in a

Any sampling plan will provide information on the quality of a lot as it is submitted—certainly more information than if no inspection is done. However, the AOQL concept is helpful in assessing the adequacy of the plan under consideration and/or indicating how it could be improved.

Ellis R. Ott
and Edward G. Schilling

lot exceed that average, as determined by the sampling plan used, then the supplier or customer undertakes 100 percent inspection. All defectives are replaced with acceptable units. Doing this brings the overall average quality for all batches or lots back up to an acceptable level. It may be, for example, that 47 out of 50 lots contain an AOQL that is acceptable to a customer based on the sampling plan used, and 3 do not. For those 3, there will be 100 percent inspection and replacement of defectives. With this replacement, the overall average quality for all 50 lots will once again be acceptable. There are tables available, based on sampling plans used, for determining the possibility of a lot not meeting an acceptable outgoing quality limit. Since 100 percent inspection is expensive and disruptive, suppliers work to keep quality up to avoid having to do this. However, 100 percent inspection is less expensive than the disruptions in a manufacturing process that may occur in accepting and using a lot of components that have more defects than expected.

> For more information, see: Eugene L. Grant and Richard S. Leavenworth, *Statistical Quality Control*, Sixth Edition (New York: McGraw-Hill, 1988); Henry L. Lefevre, *Quality Service Pays: Six Keys to Success!* (Milwaukee, WI: ASQC Quality Press, 1989).

Average Run Length (ARL) Assuming a process is in statistical control and the upper and lower control limits are known, the ARL is the expected number of subgroups inspected and measured before an operator detects whether a process has shifted to a new level of control with a new center line and control limits. There are statistical methods for calculating the ARL.

> For additional information, see: Eugene L. Grant and Richard S. Leavenworth, *Statistical Quality Control*, Sixth Edition (New York: McGraw-Hill, 1988); Donald J. Wheeler and David S. Chambers, *Statistical Process Control*, Second Edition (Knoxville, TN: SPC Press, 1992).

Bar Chart This is a widely used chart that displays magnitudes of measures or counts in horizontal or vertical bars. There are usually spaces between each bar. It is also sometimes called a *histogram.*

Managers employ bar charts to graphically make a point. For example, they can show counts of occurrences of types of events, such as different types of defects in a product. When used in this way, the bars are usually displayed from high to low. Bar charts can also display the frequency of various measurements. Figure 6 is an example of a bar chart displaying frequency. This type of graphical representation is one of the most

FIGURE 6

A standard bar chart showing the frequency of the measured length of a component in one lot.

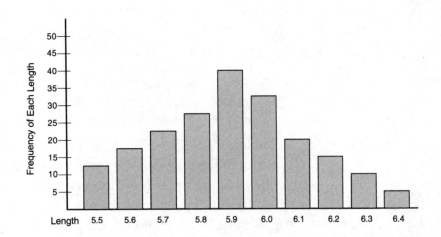

widely used tools in all work functions. Pareto charts are frequently represented as bar charts, though with no space between the bars. (See *Frequency Diagrams* and *Pareto Chart*.)

> For more information, see: Michael Brassard and Diane Ritter, *The Memory Jogger II* (Methuen, MA: GOAL/QPC, 1994); George L. Miller and LaRue L. Krumm, *The Whats, Whys & Hows of Quality Improvement* (Milwaukee, WI: ASQC Quality Press, 1992); Hy Pitt, *SPC for the Rest of Us* (Reading, MA: Addison-Wesley, 1994).

Baseline This is an initial measurement that answers the questions, what is the level of our quality, and where are we today? Employees create baselines in measurable terms and use them as a foundation for measuring future improvements.

All good process management schemes establish a baseline (for example, how the process works on an average day). Baselines then become input for the creation of new goals for process improvement. Baselines can be minimum floor-level measures of a process or of a whole system. However, more frequently, they document the current level or the average performance of a process or organization. If, for example, your car currently gets 27 miles per gallon in freeway driving, that is a baseline. After tuning the engine and following maintenance schedules, you can then measure your mileage increase against this baseline. For an organization, a baseline might be an initial Baldrige Criteria score. Thus an organization that practiced self-assessment and gave itself 450 points last year can use this to measure its progress or regression in following years. It is related to the concept of *benchmarking*. (See this entry).

> For more information, see: David A. Garvin, *Managing Quality: The Strategic and Competitive Edge* (New York: The Free Press, 1988); H. James Harrington, *Business Process Improvement* (New York: McGraw-Hill, 1991); H. James Harrington, *Total Improvement Management* (New York: McGraw-Hill, 1994).

Baseline Cycle Time This is the amount of time it takes to complete a process the first time you measure it. You can then measure all future improvements in cycle time against this baseline.

Baseline cycle time is an important concept in process improvement because most of these efforts include the goal of reducing the time it takes to complete a process. It can point to opportunities for reducing process costs while improving responsiveness to customers. To begin reducing cycle time, you must first document exactly what the existing process does and measure how much time it takes to complete. Improvements to the process would then involve reducing the baseline cycle time.

> For more information, see: George Stalk, Jr. and Thomas M. Hout, *Competing Against Time* (New York: The Free Press, 1990); Thomas H. Davenport, *Process Innovation: Reengineering Work Through Information Technology* (Boston: Harvard Business School Press, 1993).

Batch Activity This refers to a group of tasks, typically similar in nature. The act of inspecting the first of a batch is also considered a batch activity.

Bathtub Curve This is a composite curve derived from data on early failure rates and constant failure rates combined with a normalized curve of wear-out failure rates. We use this curve in assessing products that have a certain degree of reliability over time. It assumes that failures will either come early on in product life (because of product defects) or late in the product life (because it wears out). Between early and late, the rate of failure is lower and generally constant over the life of the product. Figure 7 illustrates a bathtub curve.

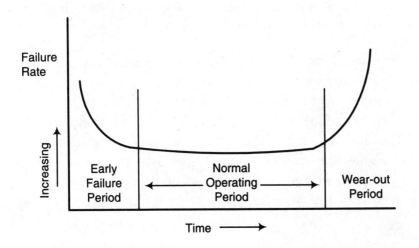

FIGURE 7
A bathtub curve.

For further information, see: Masaji Tajiri and Fumio Gotoh, *TPM Implementation* (New York: McGraw-Hill, 1992); Armand V. Feigenbaum, *Total Quality Control*, Third Edition, Revised (New York: McGraw-Hill, 1991).

Benchmark This is a quality target that answers the question, where should we be? It can be based on a number of characteristics: customer requirements, best performing comparable process, or some other characteristic. Benchmarking is also a formal process for studying and comparing one process's performance against that of another.

> In the most general terms, a benchmark was originally a sighting point from which measurements could be made or a standard against which others could be measured.
>
> *Christopher E. Bogan and Michael J. English*

Establishing benchmarks has become an integral part of most successful process improvement and reengineering efforts. Managers look for benchmarks that they or others have identified as "best-of-breed" (the best in peer industries) or "world-class" (the best in any industry) and use these as a way to evaluate their own processes and learn how to improve. Various quality groups, industry organizations, and associations now publish benchmark information for a variety of processes. Benchmarking has become very popular because it provides a concrete way to determine how well you are doing and where to improve.

> For more information, see: Gerald J. Balm, *Benchmarking: A Practitioner's Guide for Becoming and Staying Best of the Best* (Schaumburg, IL: QPMA Press, 1992); Robert C. Camp, *Benchmarking* (Milwaukee, WI: ASQC Quality Press, 1989); Gregory H. Watson, *Strategic Benchmarking* (New York: John Wiley & Sons, 1993); Christopher Bogan and Michael J. English, *Benchmarking for Best Practices* (New York: McGraw-Hill, 1994).

Benchmarking This is the act of continually comparing the performance of one process against that of another, usually of comparable or greater performance and frequently on a continuous basis. The goal of benchmarking is to identify best practices and then adapt those to improve one's own processes.

Thus, for example, a company might compare its billing process to that of a "world class" or "best-of-breed" process, such as that of American Express. Or it might compare its approach to market segmentation to that of IBM. Organizations undertake benchmarking to find out the effectiveness of their existing processes and to find ways to improve those quickly. You can use benchmarking to help determine the effectiveness of a process, the attributes and performance of a product, or just about any activity, such as the quality of a company's business strategy.

Most benchmarking projects, when properly done, follow a highly structured set of steps. These steps include (1) identifying precisely what you will benchmark, (2) creating a list of benchmarking candidates, (3) comparing data between your business activity and the benchmark, and (4) establishing goals and action plans for improvement based on what you have learned from undertaking this project.

> The key to benchmarking is to isolate common metrics in like functions (e.g., manufacturing, engineering, marketing, finance) and compare one's own business practices with those of organizations that have established themselves as leaders or innovators in that function.
>
> *Michael J. Spendolini*

While competitive analysis has always been around, a more formal process of benchmarking has become one of the most effective techniques to appear for identifying areas for improvement as well as how to improve, and by how much. Its greatest inroad has been in the area of process improvement and reengineering. Historically it has long been a technique used in engineering, manufacturing, and scientific research.

The principle behind benchmarking is that to learn how well anything works, you need something to which you can compare it. Finding outstanding companies and understanding how they perform provides guidelines for process and organizational improvement. It is useful to note that outstanding performers in a particular process do not have to be in the same industry. What is important to consider is the process itself and not necessarily the industry or company in which it is undertaken. (See also *Benchmarking, Functional*).

For more information, see: Robert C. Camp, *Benchmarking* (Milwaukee, WI: ASQC Quality Press, 1989); Robert C. Camp, *Business Process Benchmarking* (Milwaukee, WI: ASQC Quality Press, 1995); Gregory H. Watson, *Strategic Benchmarking* (New York: John Wiley & Sons, 1993); James W. Cortada and John A. Woods, *The Quality Yearbook* (New York: McGraw-Hill, 1994 and 1995);

Christopher E. Bogan and Michael J. English, *Benchmarking for Best Practices* (New York: McGraw-Hill, 1994); Robert J. Boxwell, Jr., *Benchmarking for Competitive Advantage* (New York: McGraw-Hill, 1994); Michael J. Spendolini, *The Benchmarking Book* (New York: Amacon, 1992).

Benchmarking, Competitive This is the structured process of comparing your products to those of your competition. Benchmarking topics include performance of the product, manufacturability, ease of operation, costs, maintainability, safety, and vendor support, among others.

For more information, see: Gregory H. Watson, *Strategic Benchmarking* (New York: John Wiley & Sons, 1993).

Benchmarking, Functional This is the process of comparing the operations or processes of one organization to those of one or more others. Such studies can be of a body of operations (for example, manufacturing) or a process (for example, supplier relations).

As with all benchmarking studies, the objective is to set a baseline of facts and criteria by which to measure one's own performance in comparison to that of others to gain insights leading to improved performance. Sometimes this is also called generic benchmarking. A famous example is Xerox's benchmarking of L.L. Bean's world-class warehouse inventory management processes. These processes allow L.L. Bean to consistently send out customer orders the same day the order is received. Xerox used L.L. Bean techniques for improving its own inventory management processes, though they were unrelated to shipping consumer goods. Figure 8 depicts other companies Xerox benchmarked on its way to becoming a Baldrige Award winner.

For more information, see: Gregory H. Watson, *Strategic Benchmarking* (New York: John Wiley & Sons, 1993); Robert C. Camp, *Benchmarking* (Milwaukee, WI: ASQC Quality Press, 1989); Robert C. Camp, *Business Process Benchmarking* (Milwaukee, WI: ASQC Quality Press, 1995); Christopher E. Bogan and Michael J. English, *Benchmarking for Best Practices* (New York: McGraw-

Company	Process
American Express	Collections
American Hospital Supply	Inventory control
AT&T	Research and development
Baxter International	Employee recognition and human resource management
Cummins Engine	Plant layout and design; supplier certification
Dow Chemical	Supplier certification
Florida Power & Light	The quality process
Hewlett-Packard	Research and development; engineering
L.L. Bean	Inventory control; distribution; telephonics
Marriott	Customer survey techniques
Milliken	Employee recognition
USAA	Telephonics

FIGURE 8

Selected Xerox benchmarking partners. Note that none of these are in the copier business. Adapted from *Competitive Benchmarking: The Path to a Leadership Position*, Xerox Publication 700P90261.

Hill, 1994); Robert J. Boxwell, Jr., *Benchmarking for Competitive Advantage* (New York: McGraw-Hill, 1994).

Benchmarking, Internal This is benchmarking a process or function using only benchmarking partners from within the same organization.

For more information, see: Gregory H. Watson, *Strategic Benchmarking* (New York: John Wiley & Sons, 1993); Michael J. Spendolini, *The Benchmarking Book* (New York: Amacom, 1992).

Benchmarking, Strategic This involves applying process benchmarking techniques to the level of business strategy. The purpose of this form of benchmarking is to identify strategic alternatives for how best to run a business. Managers can then choose and implement the strategy most likely to lead to an organization's success. This form of benchmarking is broader in scope than more traditional process benchmarking.

For more information, see: Gregory H. Watson, *Strategic Benchmarking* (New York: John Wiley & Sons, 1993); Christopher E. Bogan and Michael J. English, *Benchmarking for Best Practices*

(New York: McGraw-Hill, 1994); Robert C. Camp, *Business Process Benchmarking* (Milwaukee, WI: ASQC Quality Press, 1995).

Benchmarking Code of Conduct This is a set of rules of conduct governing the performance of benchmarking activities.

Benchmarking Code of Conduct

1. **Principle of Legality**. Avoid discussions or actions that might lead to or imply an interest in restraint of trade: market or customer allocation schemes; price fixing, dealing arrangements, bid rigging, bribery, or misappropriation. Do not discuss costs with competitors if costs are not elements of pricing.

2. **Principle of Exchange**. Be willing to provide the same level of information that you request in any benchmarking exchange.

3. **Principle of Confidentiality**. Treat benchmarking interchange as something confidential to the individuals and organizations involved. Information obtained must not be communicated outside the partnering organizations without prior consent of participating benchmarking partners. An organization's participation in a study should not be communicated externally without permission.

4. **Principle of Use**. Use information obtained through benchmarking partnering only for the purposes of improvement of operations within the partnering companies themselves. External use of communication of a benchmarking partner's name with its data or observed practices requires permission of that partner. Do not, as a consultant or client, extend one company's benchmarking study findings to another without the first company's permission.

5. **Principle of First Party Contact**. Initiate contacts, whenever possible, through a benchmarking contact designated by the partner company. Obtain mutual agreement with the contact on any hand-off of communication or responsibility to other parties.

6. **Principle of Third Party Contact**. Obtain an individual's permission before providing his or her name in response to a contact request.

7. **Principle of Preparation**. Demonstrate commitment to the efficiency and effectiveness of the benchmarking process with adequate preparation at each process step, particularly at initial partnering contact.

8. **Principle of Completion**. Follow through with each commitment made to your benchmarking partners in a timely manner. Complete each benchmarking study to the satisfaction of all benchmarking partners as mutually agreed.

9. **Principle of Understanding and Action**. Understand and treat your benchmarking partners as they would like to be treated. Understand how all benchmarking partners would like to have the information handled and used and do so in that manner.

FIGURE 9

The basic principles of the Benchmarking Code of Conduct. Adapted from Christopher Bogan and Michael J. English, *Benchmarking Best Practices* (New York: McGraw-Hill, 1994) and Gregory H. Watson, *Strategic Benchmarking* (New York: John Wiley & Sons, 1993).

The International Benchmarking Clearinghouse, a service of the American Productivity and Quality Center and the Strategic Planning Institute Council on Benchmarking, have adopted this Code of Conduct to promote and guide benchmarking activities among firms in the same and different industries. The code covers such issues as ownership of information, exchange of information practices, confidentiality, legality of relationships between companies, use of information, principles concerning obligations, performance, and preparation. Figure 9 summarizes the basic points in the code.

> For more information, see: Gregory H. Watson, *Strategic Benchmarking* (New York: John Wiley & Sons, 1993); Christopher E. Bogan and Michael J. English, *Benchmarking for Best Practices* (New York: McGraw-Hill, 1994); Michael J. Spendolini, *The Benchmarking Book* (New York: Amacom, 1992).

Benchmarking Gap This is the gap in performance between two processes, functions, or organizations that have been compared using a benchmarking methodology.

Gaps represent performance defects and opportunities for improvement. These can be in any activity or process you think might be improved. You might have gaps in process cycle time, cost, error rate, output quality, customer satisfaction, and so on.

> For more information, see: Gregory H. Watson, *Strategic Benchmarking* (New York: John Wiley & Sons, 1993); Christopher E. Bogan and Michael J. English, *Benchmarking for Best Practices* (New York: McGraw-Hill, 1994).

Benchmarking Partner This is the company or persons you work with to conduct a benchmarking study. The phrase also refers to two or more companies working together to compare similar functions to learn how both can improve their operations.

Relationships among partners are usually open and sharing in describing operations. There is close cooperation to help each other learn about how well they are doing and where they might improve. The partners usually have an agreement defining their relationship. One trend is the creation of consortiums for benchmarking. Many companies will come together to benchmark a specific process or best practices within a functional area (for example, information systems). Major management consulting companies with specialized skills also sponsor similar benchmarking projects.

> For more information, see: Gregory H. Watson, *Strategic Benchmarking* (New York: John Wiley & Sons, 1993); Michael J. Spendolini, *The Benchmarking Book* (New York: Amacom, 1992). A useful organization to contact for information and help is the American Productivity and Quality Center, 123 N. Post Oak Lane, Suite 300, Houston, TX 77024-7797, USA; telephone: (800) 776-9676. This organization administers an annual benchmarking award.

Best-of-Breed This term refers to the best example of a process or measurement available, usually within an industry. Sometimes the term refers to the best worldwide regardless of industry.

The concept has become important in the past several years largely in response to competitive pressures to improve the quality of goods and services. There has been a significant increase in the application of comparative techniques (for example, benchmarking) to identify best practices. Primary concern has been among competitors to identify best practices within an industry. A growing body of knowledge regarding best-of-breed has been developing across most functional areas, major processes, and industries.

> For more information, see: Gregory H. Watson, *Strategic Benchmarking* (New York: John Wiley & Sons, 1993); Robert J. Boxwell, Jr., *Benchmarking for Competitive Advantage* (New York: McGraw-Hill, 1994).

Best-of-Class This is another way of saying best-of-breed, superior to all comparable goods or services in the same industry.

Best Practice This means outstanding performance within an activity or process regardless of industry, like the term world class.

Best practices are considered to be more creative and effective than similar practices in other companies. They are thoroughly documented, well-measured, and effectively managed based on fact gathering and analysis. They yield better outcomes (higher quality at lower costs) than comparable procedures.

Best practices also tend to receive wide attention (for example, billing at American Express, logistics at Federal Express, the employee suggestion process at Milliken). Managers usually identify and recognize best practices as a by-product of formal benchmarking.

Best practices has legitimized plagiarism.

Jack Welch

> For more information, see: Alexander Hiam, *Closing the Quality Gap* (Englewood Cliffs, NJ: Prentice Hall, 1992); Gerald J. Balm, *Benchmarking: A Practitioner's Guide for Becoming and Staying Best of the Best* (Schaumburg, IL: QPMA Press, 1992); Christopher E. Bogan and Michael J. English, *Benchmarking for Best Practices* (New York: McGraw-Hill, 1994).

Bias This is any tendency of a process or human outlook to be a source of error in the perception of performance. More technically, it is similar to accuracy: It is the systematic difference between a true value and the average of the distribution of values.

In processes, a difficult-to-find, erratic special cause can create a bias in the process output or in the measurement of that output. A goal of quality control and improvement is to remove bias from processes and their measurement.

> For more information, see: Joseph M. Juran and Frank M. Gryna, *Quality Planning and Analysis*, Third Edition, (New York: McGraw-Hill 1993).

Big C, little c This phrase contrasts differences between customers who pay for services and goods (Big C or external customers), and fellow employees (little c or internal customers).

The concept is useful in process improvement and reengineering. It reminds us that all people can be viewed as customers and that we should treat them with courtesy and give them the best quality goods and services possible. This is particularly the case with little c—fellow employees or business partners—who, within a process, depend on the performance of other employees to do their own work well. Everyone in an organization should look to delighting their little c customers with the same attention as to their Big C Customers who buy the finished product or service.

For more information, see: Jack D. Orsburn et al., *Self-Directed Work Teams: The New American Challenge* (Burr Ridge, IL: Irwin Professional Publishing, 1990); Jerry Bowles and Joshua Hammond, *Beyond Quality: How 50 Winning Companies Use Continuous Improvement* (New York: G.P. Putnam's Sons, 1991); Richard J. Schonberger, *Building a Chain of Customers* (New York: The Free Press, 1990).

Big Q, little q This phrase contrasts differences between managing for quality in a global sense, for example in business products and processes, and managing quality in a more narrowly defined scope, such as for one product or process. TQM recognizes that little q is fundamentally the result of the implementation of Big Q attitudes, values, and practices in the company's culture and its operations. Figure 10 provides several distinctions between Big Q and little q.

For more information, see: H.G. Menon, *TQM in New Product Manufacturing* (New York: McGraw-Hill, 1992); Joseph M. Juran, *Juran on Leadership for Quality: An Executive Handbook* (New York: The Free Press, 1989).

Bimodal Distribution This is a distribution with two modes in the frequency of occurrence of whatever population is being measured. (The mode is the most frequently occurring value in

FIGURE 10

Contrasting Big Q and little q. Reproduced from *Making Quality Happen* with permission of the copyright holder, Juran Institute, Inc. All rights reserved.

TOPIC	CONTENT OF BIG Q	CONTENT OF little q
Products	All products, goods, and services, whether for sale or not	Manufactured goods
Processes	All processes; manufacturing support, business, etc.	Processes directly related to manufacture of goods
Functions	All functions	Those directly associated with manufacture of goods
Facilities	All facilities	Factories
Industries	All industries; manufacturing, service, government, etc., whether for profit or not	Manufacturing
Cost of Poor Quality	All costs that would disappear if everything were perfect	Costs associated with deficient manufactured goods

a data set.) It is usually displayed using a histogram (see *Histogram* for more on this tool).

A bimodal distribution documents a relationship between two factors in a process, such as time of day and level of traffic. Such a representation can lead you to ask the obvious question: Why is there a difference? From this, we can gain a better understanding of processes, their operation, and their improvement. Figure 11A illustrates a bimodal distribution for the number of pizzas delivered during an average day for a pizza restaurant. With this information, a manager could better plan staffing decisions.

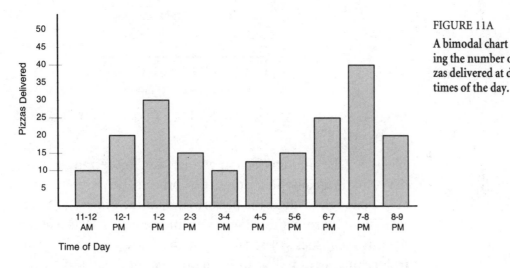

FIGURE 11A

A bimodal chart showing the number of pizzas delivered at different times of the day.

Sometimes two sets of data are combined and they create a bimodal distribution. An example would be to look at the error rates of workers on the day shift and the night shift. By combining these curves, it may become obvious that the night shift has a higher error rate than the day shift. You can then begin to explore why this is happening and what you can do about it. Figure 11B shows a bimodal distribution used this way.

For more information, see: Henry L. Lefevre, *Quality Service Pays: Six Keys to Success!* (Milwaukee, WI: ASQC Quality Press, 1989); George L. Miller and LaRue L. Krumm, *The Whats, Whys & Hows of Quality Improvement* (Milwaukee, WI: ASQC Quality Press,

FIGURE 11B

This is another form of bimodal distribution in which the distributions for two groups have been combined.

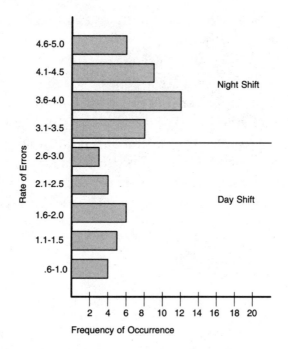

1992); Hy Pitt, *SPC for the Rest of Us* (Reading, MA: Addison-Wesley, 1994).

Blemish Like an imperfection on our skin, this term refers to a deviation in quality in a product or process. However, it is not severe enough to affect the performance of the process or product. These are sometimes known as "incidental" defects. We can identify blemishes statistically through a variety of quality control techniques as well as through observation.

For more information, see: Armand V. Feigenbaum, *Total Quality Control*, Third Edition, Revised (New York: McGraw-Hill, 1991).

Blind Survey This is a survey of customers or other individuals in which the survey's sponsor and the respondents are unknown to each other. Companies often use this technique to compare the performance of one company or process against another. Blind surveys have well-defined steps for their execution. They help eliminate bias in responses and are a valuable tool for understanding consumer issues and competitors' standings in the marketplace.

Companies undertake surveys to learn about people's knowledge, beliefs, preferences, satisfaction, and so on.

Philip Kotler

For more information, see: Bob E. Hayes, *Measuring Customer Satisfaction: Development and Use of Questionnaires* (Milwaukee, WI: ASQC Quality Press, 1992); M. Hannan and P. Karp, *Customer Satisfaction: How to Maximize, Measure, and Market Your Company's "Ultimate Product"* (New York: Amacon, 1989).

Block Diagram A block diagram is a flowchart illustrating the steps, interdependencies, and operational relationships in a process. The name comes from the fact that the diagram shows process steps as a series of geometric block-shaped figures.

Block diagrams illustrate the flow of work and materials through a process, how the steps relate to each other, and where the process begins and ends. Figure 12 illustrates a block diagram for changing the oil in a car.

For more information, see: H. James Harrington, *Business Process Improvement* (New York: McGraw-Hill, 1991); George L. Miller and LaRue L. Krumm, *The Whats, Whys & Hows of Quality Improvement* (Milwaukee, WI: ASQC Quality Press, 1992).

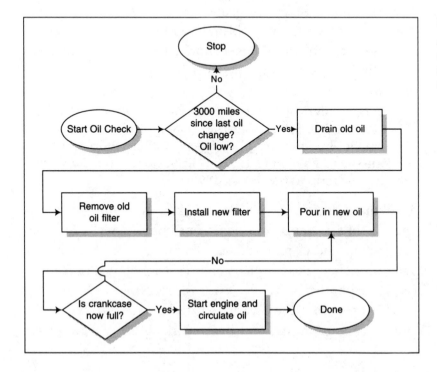

FIGURE 12

A simple block chart showing the flow of steps in changing oil in a car.

Brainstorming This is a widely used technique for a group to generate many ideas in a short time, often to improve a process or solve a problem.

A technique for brainstorming includes the following steps:

- State the subject for discussion.
- Give all participants a chance to think about the subject for a few minutes.
- Review the rules of the session, such as: Everyone must participate; no discussion of ideas; write down every idea on a flip chart without evaluation.
- Determine whether ideas will be offered as they come to people or by systematically going around the table.
- Conduct the session.

A successful brainstorm lets people be as creative as possible and does not restrict their ideas in any way. This free-form approach can generate excitement in the group, equalize involvement, and often result in original solutions to problems.

Peter R. Scholtes

After all the ideas are offered, use a technique like multivoting or nominal group technique to come to a consensus around one or more ideas to implement. One technique to spark imagination—a critical component for a successful brainstorming exercise—is to use a mind game that takes participants away from thinking about the problem in conventional terms. A popular tool for doing this is Roger von Oech's *Creative Whack Pack* (Stamford, CT: U.S. Games Systems, Inc., 1992). (See also *Multivoting.*)

For more information, see: Peter R. Scholtes, *The Team Handbook* (Madison, WI: Joiner Associates, 1988); Bruce Brocka and M. Suzanne Brocka, *Quality Management,* (Burr Ridge, IL: Irwin Professional Publishing, 1992); William Lareau, *American Samurai* (New York: Warner Books, 1991).

Breakthrough Often also called breakthrough thinking, this is the idea of searching for a significant improvement in the design, set up, and/or performance of a process. Emphasis is on a radically new, or revolutionary approach as opposed to the evolutionary improvement in an existing process.

Breakthrough means change, a dynamic, decisive movement to new, higher levels of performance.

Joseph M. Juran

Managers may use a variety of techniques to facilitate the process; brainstorming and benchmarking are the two most popular. Other techniques involve rigorous analysis of a process's attributes, with extensive reliance on data gathering and statistical process control (SPC) methods. This approach helps managers gain insight into deep problems in a process and how they might change and improve it. The application of SPC methods, for example, is at the heart of the arguments made by such quality gurus as W. Edwards Deming, Kaoru Ishikawa, Armand V. Feigenbaum, and Joseph M. Juran.

> For more information, see: Gerald Nadler and Shozo Hibino, *Breakthrough Thinking* (Rocklin, CA: Prima Publishing, 1990); Joseph M. Juran, *Managerial Breakthrough,* Revised Edition (New York: McGraw-Hill, 1994); Peter R. Scholtes, *The Team Handbook* (Madison, WI: Joiner Associates, 1988); Michael Hammer and James Champy, *Reengineering the Corporation* (New York: HarperBusiness, 1993); Peter M. Senge, *The Fifth Discipline* (New York: Doubleday Currency, 1990).

Breakthrough Action Teams (BATs) This is a name for process improvement or process reengineering teams. Such teams, when working well, develop original suggestions for change that have high leverage for substantially improving an entire process.

> For more information, see: James W. Cortada and John A. Woods, *The Quality Yearbook* (New York: McGraw-Hill, published annually).

Bug This slang term refers to a problem or error that negatively affects operational performance. The term originated in the information processing industry during the 1940s and early 1950s, where it has always been used to describe a programming error. People also use this word to describe problems in processes that they have not yet identified. It is said to have originated with Admiral Grace Hopper, a pioneer in data processing, when she found a moth in a computer.

> For more information, see: Paul F. Wilson, Larry D. Dell, and Gaylord F. Anderson, *Root Cause Analysis* (Milwaukee, WI: ASQC Quality Press, 1993).

Burn-in This is the act of final stress testing of a procedure, process, machine, or software in live operating conditions. This helps operators find the last remaining problems before putting the item into daily use.

For more information, see: Armand V. Feigenbaum, *Total Quality Control*, Third Edition, Revised (New York: McGraw-Hill, 1991).

Business Process This is a process used in an office environment, such as document preparation or customer service phone systems. This term distinguishes these processes from other types, such as those used in manufacturing.

C Chart This is a count chart of any type (see *Count Chart*).

Calibration In quality management, this is a comparison of one measurement system or instrument not verified as accurate to another standard measurement system or instrument with verified accuracy. This is done to identify variations from a required specification for performance.

Calibration is often used in engineering and manufacturing environments that require multiple comparisons by different means to ensure the accuracy of specific data.

> For more information, see: James Robert Taylor, *Quality Control Systems* (New York: McGraw-Hill, 1989); H.G. Menon, *TQM in New Product Manufacturing* (New York: McGraw-Hill, 1992); Armand V. Feigenbaum, *Total Quality Control*, Third Edition, Revised (New York: McGraw-Hill, 1991).

Capability Process Index (Cp) This is a technique for making a ratio of the specification width (that is, the range within which an output is considered to meet specifications) in relation to the process's natural distribution for assessing whether the process can regularly meet specifications.

It is primarily used in manufacturing and indicates the potential of a process to produce goods or services within specifications. Managers use process capability to (1) determine which

One does what one can, not what one can't.

Attributed to Agatha Christie

processes are not capable of meeting specifications, (2) to identify those processes operating suboptimally, (3) to estimate the output proportion that will not conform to specifications, and (4) to evaluate process performance over extended periods to maintain optimal operating practices.

Process capability is calculated by taking the difference between the upper specification limit (USL) and the lower specification limit (LSL) divided by the 6 times the standard deviation (σ or sigma) of the process (USL–LSL/6σ). Why 6σ? Because 3 standard deviations on either side of the center line of the control chart mark the upper and lower control limits of the process. The goal in this calculation, thus, is to see if the specification limits fall inside or outside the process control limits. If the solution to this formula is 1, this means the specification limits and the control limits are equal.

For example, if the USL for gas mileage on a car is 20 mpg and the LSL is 16 mpg and the standard deviation is .6, then the calculation for the process capability is 4/3.6, which equals 1.11. A process capability index over 1 means the process (that is, the car's engine in this case) is capable of meeting specifications within the range of 6 standard deviations. If the index is less than 1, this means that process outputs will often not meet specifications. If the process cannot meet specifications, you need to either improve the process or change the specifications.

The goal is to get the process up into the area of 2 or higher. The higher the process capability index, the more reliable your process. Many American firms strive for Cp of 1, while Toyota Motors, for example, frequently obtains process capability indices of 2.0. A Cp of greater than 1 indicates the natural variability of the process is less that the specification range. A higher number for the capability index is desirable because it indicates variation within the process is much less than the specification limits. It means outputs are more likely to meet target specifications. Figure 13 shows the standard distribution of a process with the specification limits falling inside the control limits. Thus the Cp is less than 1.

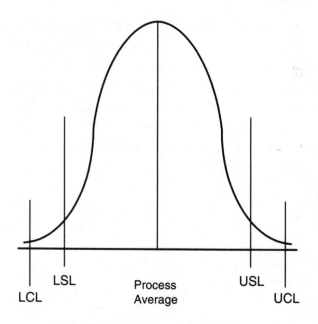

LSL   Process   USL
LCL   Average   UCL

FIGURE 13
This figure shows the capability of a process with the specification limits falling inside the process limits. This means that the process will deliver some percent of output that does not meet specifications.

For more information, see: George L. Miller and LaRue L. Krumm, *The Whats, Whys & Hows of Quality Improvement* (Milwaukee, WI: ASQC Press, 1992); John L. Hradesky, *Productivity and Quality Management* (New York: McGraw-Hill, 1988); Hy Pitt, *SPC for the Rest of Us* (Reading, MA: Addison-Wesley, 1994); and many other books on quality management statistical tools.

Capability Process Index (Cpk) This is another type of capability index based on the actual performance of the system. This is calculated by taking the mean of a process performance and using it to understand the capability of the system in terms of specification versus *actual performance.*

In this case, we are interested in how well the process is performing at either the upper or lower end of the specification limit. Thus the calculation is slightly different from that for Cp. In terms of our previous example, let's say the actual mean gas mileage for the car was 17 mpg. We would then perform the following calculation: $|LSL—\mu|/3\sigma$. (The vertical lines mean that the quantity within them is always positive. μ = mean. We use 3σ because we are only looking at the process's mean in terms of its closeness to either the upper or lower specification limit.) In our problem, we would have |16–17|/1.8. This would equal 1/1.8 or .56. The goal is for the actual mean performance to mirror the hypothetical mean for the specification (in this case 18, but actual performance only gave us 17). A Cpk this much below 1 means the process is not performing well and that the company needs to review this to determine what is causing this kind of skewed variation. Figure 14 illustrates this problem.

FIGURE 14

Cpk for car with LSL of 16 mpg and USL of 20 mpg that performed at 17 mpg.

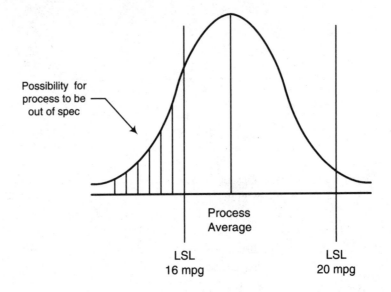

Possibility for process to be out of spec

Process Average

LSL
16 mpg

LSL
20 mpg

For more information, see: George L. Miller and LaRue L. Krumm, *The Whats, Whys & Hows of Quality Improvement* (Milwaukee, WI: ASQC Quality Press, 1992); John L. Hradesky, *Productivity and Quality Management* (New York: McGraw-Hill, 1988); Hy Pitt, *SPC for the Rest of Us* (Reading, MA: Addison-Wesley, 1994).

Carry Over This distinguishes features from an old process, product, or service incorporated into a new process, product, or service.

When reengineering, for example, there are practices from the former process that will make sense to incorporate into the new scheme. We do this based on experience with what works well and what does not. For instance, say you are centralizing and computerizing the billing function formerly performed manually in dozens of offices. You might want to carry over features dealing with record keeping and accounting conventions (but now automated) or policies and criteria for various write-offs. The same applies to products; for example, automobile manufacturers often carry over from one model to the next various components that work well, such as an automatic transmission.

Catastrophic Failure This is considered a sudden, complete, and unpredictable failure of a process or system. "Crash" is another term used to describe the same crisis. It is a phrase drawn from information processing.

An example of a catastrophic failure would be a power shutdown due to the sudden failure of some key component or process. Catastrophic failures are always the result of special causes.

Causal Analysis This technique concentrates on (1) identifying specific failures or defects, (2) discovering root causes of those failures, and (3) concluding with recommendations on how to eliminate those defects by implementing solutions that address the appropriate cause. It is a popular component of defect prevention activities within a process.

> For more information, see: James H. Saylor, *TQM Field Manual* (New York: McGraw-Hill, 1992); George L. Miller and LaRue L. Krumm, *The Whats, Whys & Hows of Quality Improvement* (Milwaukee, WI: ASQC Quality Press, 1992).

Cause This is a reason for a specific condition, action, or effect.

In the world of process improvement and reengineering, identifying and addressing a cause is a serious subject, approached through a variety of formal techniques. This is a fundamental concept for undertaking process improvement or applying quality management tools. The successful use of statistical process control methods and Total Quality Management tools relies on a solid understanding of the causes of variation, problems, defects, and other unpredictable behaviors that affect productivity and the failure to meet customer requirements. Personal experiences and guesswork, on the other hand, are not sound ways to understand the causes of a problem.

Successful process improvement activities involve the careful identification of causes of problems. Managers, engineers, and line workers use such tools as run and control charts, benchmarking, cause-and-effect diagrams, root cause analysis, and a variety of statistical process control techniques and formulas to identify causes.

For more information, see: Paul F. Wilson, Larry D. Dell, and Gaylord F. Anderson, *Root Cause Analysis* (Milwaukee, WI: ASQC Quality Press, 1993); James H. Saylor, *TQM Field Manual* (New York: McGraw-Hill, 1992); N. Logothetis, *Managing For Total Quality: From Deming to Taguchi and SPC* (Englewood Cliffs, NJ: Prentice Hall, 1992); George L. Miller and LaRue L. Krumm, *The Whats, Whys & Hows of Quality Improvement* (Milwaukee, WI: ASQC Quality Press, 1992).

Cause-and-Effect Analysis This is a method used to display and study underlying causes. Some people call this "fishboning" since the output is a chart that takes the shape of a fish skeleton. The analysis was first proposed by Kaoru Ishikawa, a Japanese pioneer in quality control methodologies.

In undertaking this analysis, you write down an effect (usually a problem, or problem statement as it is sometimes called); for example, the kitchen workers are overcooking pizzas. You then identify the possible causes of that effect using a cause-and-effect diagram like that shown in figure 15. As the figure suggests, there are usually five quadrants to the diagram; each identifies possible root causes in one of five major categories ("5 Ms"): machinery, methods, manpower (human resources), measurements, and materials. (Sometimes you will see other names for these categories, such as process for methods.) Once you have completed the diagram, you look at the breakdown of categories to identify a root cause or causes that seem responsible for the effect. In the pizza case, it could be poor training of employees. The overcooked pizza is a symptom of this deeper problem concerning training. This could include training on how to cook pizza, operate the oven, or properly combine the ingredients. This same analysis can also help you understand why a process is delivering quality outputs. Figure 15 illustrates the generic format for a cause-and-effect or fishbone diagram.

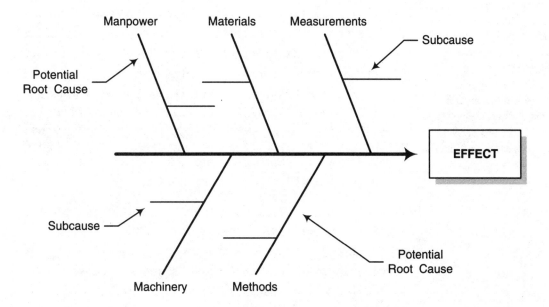

For more information, see: Paul F. Wilson, Larry D. Dell, and Gaylord F. Anderson, *Root Cause Analysis* (Milwaukee, WI: ASQC Quality Press, 1993); James H. Saylor, *TQM Field Manual* (New York: McGraw-Hill, 1992); Kaoru Ishikawa, *Introduction to Quality Control* (Tokyo, Japan: 3A Corporation, 1990); John L. Hradesky, *Productivity and Quality Improvement* (New York: McGraw-Hill, 1988). Most books on quality management techniques include coverage of cause-and-effect analysis.

FIGURE 15

A generic version of a cause-and-effect diagram. The lines coming from the spine are root causes and the smaller lines are subcauses.

Cause-and-Effect Diagram This type of graph illustrates the relationship between one problem and its possible various causes. People popularly call it a fishbone diagram because it looks like the skeleton of a fish with each "bone" providing a possible answer to the question *why?* This is a standard quality tool for identifying and dealing with process problems.

See figure 15 in the definition for cause-and-effect analysis just above for an example of the diagram. It is also known as an "Ishikawa diagram," after Kaoru Ishikawa, who first popularized the tool. He is also known for developing the concept of quality circles in the early 1960s.

For more information: See references for *Cause-and-Effect Analysis.*

Central Line This is the line on a control chart that represents the long-term average or mean of the measurement being charted. The lower control limit falls below the central line, and the upper control limit above the central line.

For more information, see: Hy Pitt, *SPC for the Rest of Us* (Reading, MA: Addison-Wesley, 1994); George L. Miller and LaRue L. Krumm, *The Whats, Whys and Hows of Quality Improvement* (Milwaukee, WI: ASQC Quality Press, 1992); Brian L. Joiner, *Fourth Generation Management* (New York: McGraw-Hill, 1994); and any reference that covers control charts.

Central Tendency This is any tendency of data from a process to cluster together around one or more values at a point between high and low values for that process.

For example, in the distribution of school grades, we can say that in a class where most students earn Bs on test after test, this shows a central tendency to cluster around a B instead of spread out with As, Bs, Cs, Ds and Fs in a more even distribution. The most common measures of central tendency in statistics are the *median,* the *mode,* and the arithmetic *mean.*

Certainty This word describes a degree of probability. It is most often used to define the relative capability of measuring a quantity. It answers the question: "How certain are we of measuring something accurately?" The answer might be "with 95 percent certainty," which says that there is a 5 percent chance that the measurement will be wrong.

Certificate of Compliance This kind of document is issued by a company, purchasing agent, or quality control department to certify that a supplier has complied with required levels of qual-

When used in conjunction with other statistical tools such as Pareto diagrams, cause-and-effect diagrams are useful for promoting process improvement on a priority basis, accumulating and organzing knowledge and technology, consolidating the ideas of all employees for control-related activities, and facilitating discussions, education, and a variety of other aspects of human relations.

Kaoru Ishikawa

ity, service, conformance to requirements, or regulations. This document is often found in companies that have formal quality control processes, such as ISO 9000 certified manufacturing locations.

> For more information, see: James L. Lamprecht, *ISO 9000: Preparing for Registration* (Milwaukee, WI: ASQC Quality Press, 1992); Robert W. Peach, *The ISO 9000 Handbook*, Second Edition (Milwaukee, WI: ASQC Quality Press, 1994); John T. Rabbit and Peter A. Bergh, *The ISO 9000 Book*, Second Edition, (New York: Amacom, 1994); Eugenia K. Brumm, *Managing Records for ISO 9000 Compliance* (Milwaukee, WI: ASQC Quality Press, 1995).

Certification The process by which an individual or organization is certified by a third party as having a certain level of systems, skills, or capabilities to deliver certain levels and types of services.

Certification of quality experts as provided, for example, by the ASQC, has long been a common practice in manufacturing. Certification by an ISO-authorized auditor that a manufacturer has achieved an ISO 9000 level of process documentation is another example. Companies moving toward employee compensation and reward processes that emphasize skill development often use certification processes as a way of validating achievement.

> For more information, see: James L. Lamprecht, *ISO 9000: Preparing for Registration* (Milwaukee, WI: ASQC Quality Press, 1992); Robert W. Peach, *The ISO 9000 Handbook*, Second Edition (Milwaukee, WI: ASQC Quality Press, 1994); John T. Rabbit and Peter A. Bergh, *The ISO 9000 Book*, Second Edition, (New York: Amacom, 1994); Eugenia K. Brumm, *Managing Records for ISO 9000 Compliance* (Milwaukee, WI: ASQC Quality Press, 1995).

Certification Body This is a third-party organization that certifies the skill of an individual or organization to perform specific services. Regulatory agencies can serve as certification bodies but so do ISO bodies, such as the American Society for Quality Control, which certifies the skills of individuals. Even quality organizations within a company (such as a manufacturing division) may have procedures for certifying its employees in various skills and job activities.

Every major industrialized country in the world today has a variety of ISO 9000 certification organizations. Those nations that have national quality awards also have a variety of quality certification bodies. Certification bodies are considered to be unbiased, competent, and are expected to conduct themselves with absolute integrity.

Chain Sampling Plan This is a type of acceptance sampling by which operators use acceptance criteria that depend partially on results of immediately preceding lots in inspecting new samples.

Harold F. Dodge, who developed this plan, suggests that chain sampling is used under these conditions: (1) interest centers on a single quality characteristic that involves costly or destructive tests, such that only a small number of tests per lot can be justified; (2) the inspected product is from a series of successive lots manufactured by an essentially continuing process; (3) under normal conditions, the lots are expected to be essentially the same quality; and (4) the lot comes from a source in which the customer has confidence. Under these conditions, if samples from the first lot are found to have zero defects, then one can expect that the succeeding lot will have zero and not more than one defect.

> For more information, see: Eugene L. Grant and Richard S. Leavenworth, *Statistical Quality Control*, Sixth Edition (New York: McGraw-Hill, 1988).

Champion In quality management this refers to an individual who is responsible for implementing or supporting quality improvement efforts within an enterprise.

In recent years people have used this word to designate owners of specific processes or improvement efforts beyond their normal duties. Thus, a vice president of sales may also be the champion of quality for a customer order service process. This term also has been used in a program at Bell Atlantic, which emphasized the idea of identifying corporate entrepreneurs, training them, and developing their ideas into busi-

A "Quality First" champion—an individual that leads by his or her own participation—is needed to inspire others to participate fully in the change process.

V. Daniel Hunt

nesses. After a rigorous evaluation of proposals by would-be champions, those individuals whose ideas pass muster become the leaders of the teams that bring the products or services to market.

> For more information, see: V. Daniel Hunt, *Quality in America* (Burr Ridge, IL: Irwin Professional Publishing, 1992); on Bell Atlantic's program, Alexander Hiam, *Closing the Quality Gap* (Englewood Cliffs, NJ: Prentice Hall, 1992).

Chance Causes This is another term for common causes of variation within a process. These are causes small and difficult to identify, but can still influence the quality of performance of a process. The sources of chance variation include: the environment, inspection, material variations, personnel, and processes.

Any variation in a process due to common causes can be addressed only by learning more about the process as a whole and the system of which it is a part. Then by experimenting with the system to make adjustments in how all the pieces interact, you can make changes that will reduce that variation. This is what quality management teaches. We can contrast common causes with special causes, which are identifiable sources that result in process outputs exceeding control limits. Managers can immediately identify such causes and eliminate them to bring the process back into control. The problem is that traditional managers often treat common cause variation as if it were due to special causes. This is what Deming calls "tampering," which simply adds one more element to the system that can increase rather than reduce variation. (See also *Special Causes* and *Tampering*.)

> For more information, see: George L. Miller and LaRue L. Krumm, *The Whats, Whys & Hows of Quality Improvement* (Milwaukee, WI: ASQC Quality Press, 1992).

Chance Variation This type of variation is the result of chance or common causes. A goal of statistical process control is to better understand the level of such variation and then, through experiment and other improvement methodologies, to improve the process to reduce chance variation.

> For more information, see: George L. Miller and LaRue L. Krumm, *The Whats, Whys & Hows of Quality Improvement* (Milwaukee, WI: ASQC Quality Press, 1992).

Change Control This is the process used to make changes in products, processes, or organizations. It is usually a documented plan or an organized process designed to implement desired change while minimizing the possibility of unexpected and unwanted changes. A change control is valuable when modifying software, for example. It allows the predictable management of changes.

Some commonly used change control procedures occur in product design and development, software programming and maintenance, process management, and, increasingly, in the cultural transformation of organizations.

Change Management This is the same as change control; however, some people also use it to describe what an organization must do to change its culture to conform to quality practices.

Managing a culture change can include (1) measuring the ability of an organization to change, (2) determining the degree of change it can accept, and (3) developing strategies, training programs, measurement systems, and roadmaps to bring the change about.

> For more information, see: James W. Cortada, *TQM for Sales and Marketing Management* (New York: McGraw-Hill, 1993); James W. Cortada and John A. Woods, *The Quality Yearbook* (New York: McGraw-Hill, 1994, 1995); H. James Harrington, *Business Process Improvement* (New York: McGraw-Hill, 1991).

Chart This can be (1) a graphical representation of data showing significant trends and relationships or (2) a basic tool for statistically documenting performance of a process. As a hammer and saw are to a carpenter, so a chart is to anyone working with data to understand processes. In TQM, charts document the current state of affairs and provide insight into where improvements might be made.

When we refer to the seven tools of Total Quality Management, we are talking about a variety of charts. Widely used charts include Pareto diagrams, checklists, checksheets, cause-and-effect diagrams, histograms, control charts, and scatter plots. All share one thing in common: They portray data in graphical format, often displaying these in time sequence. It is important that data displayed on charts be gathered in a structured and consistent way.

> For more information, see: Michael Brassard and Diane Ritter, *The Memory Jogger II* (Methuen, MA: GOAL/QPC, 1994); Hy Pitt, *SPC for the Rest of Us* (Reading, MA: Addison-Wesley, 1994); PQ Systems, Inc., *Total Quality Transformation Improvement Tools* (Miamisburg, OH: Productivity Quality Systems, Inc., 1994, 800-777-3020); Nancy R. Tague, *The Quality Toolbox* (Milwaukee, WI: ASQC Quality Press, 1995); and any of several books that include coverage of basic quality improvement tools.

Charter This usually refers to the list of tasks a project team is authorized to perform. It can also refer to the mandate or responsibility of a particular department, division, or group to execute and manage a particular set of activities for the entire organization.

For example, in sales, a particular division might have responsibility—the charter—to design, sell, and service a particular product to the exclusion of all other divisions within the company. Charters of this type are always announced and enforced to reduce replication of services or competition within the enterprise, both of which are counterproductive and expensive.

> For more information, see: Kimball Fisher et al., *Tips for Teams* (New York: McGraw-Hill, 1995); Joseph M. Juran, *Juran on Leadership for Quality* (New York: The Free Press, 1989); Peter R. Scholtes, *The Team Handbook* (Madison, WI: Joiner Associates, 1988).

Checklist This refers to a list of steps or actions in a process that are implemented and checked off from a special list as completed; sometimes it takes the form of a "to do" list.

A team charter provides a sense of purpose, as well as a clear definition of the team's role and expectations.

From Tips for Teams

It is the most rudimentary form of project management documentation. It may include a list of projects and processes to check off when completed. It may include an activity list with the names of those responsible for completing each one and by when. It may also be a list of problems to help determine which kinds happen most frequently. The inclusion of checklists as a quality tool indicates that managing for quality is not a set of exotic new practices. What TQM requires, though, is the disciplined and informed application of such tools. This approach allows us to become more efficient and effective at making processes and systems run well and provides the data for continuously improving them. Figure 16 illustrates a checklist for the preparation of the manuscript for this book.

For more information, see: any of several books that include coverage of basic quality improvement tools.

FIGURE 16

A standard checklist. As each task is completed, the user checks it off.

CHECKLIST FOR PREPARATION OF MANUSCRIPT

TASKS	DONE
Step 1. Collect Terms for Inclusion	✓
Step 2. Write Definitions	✓
Step 3. Provide References	✓
Step 4. Develop Figures	✓
Step 5. Have Experts Review First Draft	
Step 6. Revise Manuscript	
Step 7. Deliver Manuscript to Publisher	

Checksheet This is a tool for collecting data in a consistent fashion. It is used to efficiently record counts of events or attributes.

A checksheet is set up so each person collecting data will record it the same way. Because consistency in data collection is important, the design of a checksheet is also important. The goal is to make it easy to use and to avoid confusion and ambiguity. The most common checksheets take a matrix format, with columns and rows for recording hash marks for the number of events, defects, or whatever type data people are collecting. Figure 17 shows a common checksheet format.

FIGURE 17

A standard version of a checksheet. This tool helps in counting various events or attributes.

COMMON CAR PROBLEMS CHECKSHEET

PROBLEM	JAN	FEB	MAR	APR	TOTALS
Clutch Slippage	卌	卌 II	卌	卌 III	25
Brakes	卌卌	卌卌卌I	卌卌卌I	卌 II	39
Radiator Leaks	卌 III	卌 I	卌	卌 IIII	27
Air Conditioner Problems	卌IIII	卌 II	卌 II	卌卌	33
Oil Leaks	卌卌	卌卌	卌卌	卌 IIII	39

A second checksheet often used is for identifying the location of defects on objects going through a process. This type of checksheet is usually a simple drawing of the item being checked. The person collecting the data keeps a tally of the location of different defects on the drawing. If several defects turn up in one location on an item, this helps identify potential causes of the defect. Figure 18 (page 66) illustrates a defect location checksheet.

> For more information, see: PQ Systems, Inc., *Total Quality Transformation Improvement Tools* (Miamisburg, OH: Productivity Quality Systems, Inc., 1994; 800-777-3020); Michael Brassard and Diane Ritter, *The Memory Jogger II* (Methuen, MA: Goal/QPC, 1994); Nancy R. Tague, *The Quality Toolbox* (Milwaukee, WI: ASQC Quality Press, 1995); and any of several books that include coverage of basic quality improvement tools.

Chronic Losses These are losses caused by an ongoing quality deficiency that is inherent in a system, process, service, or product. The gap between a piece of equipment's actual capability and its optimal capability is also referred to as chronic loss when the same loss occurs repeatedly within a small range of incidence.

> For more information, see: Masaji Tajiri and Fumio Gotoh, *TPM Implementation* (New York: McGraw-Hill, 1992).

FIGURE 18

A checksheet showing
the location of defects
on a car door.

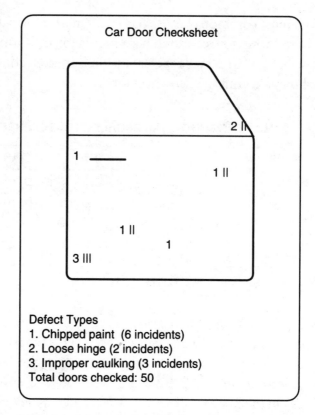

FIGURE 18

A checksheet showing
the location of defects
on a car door.

Cloning This is a biological metaphor for the application of a successful solution, remedy, or process to multiple products or processes.

Cloning is a commonly used word in manufacturing. It often refers to the design and manufacture of a product that looks and functions very much like a competitor's. In quality management, it is the replication of a stabilized process that is within control limits that we can now apply elsewhere within an enterprise.

Closed-Loop Process This is a process that includes as part of itself a device for sensing and measuring feedback and making automatic adjustments in response to this feedback. The cooling system in an automobile that includes a thermostat is an example of a closed-loop process. This device, as part of the system, keeps the engine operating within a narrow temperature range.

This idea has crept into process improvement activities in the past several years. The goal has been to institutionalize improvement by incorporating feedback mechanisms as part of all organizational processes to keep them stable and operating within control limits. Using feedback in the form of statistical process control and other charts, and information from customers, individuals and teams in these organizations have the responsibility and authority to keep the process in control, identify and deal with special causes, and make improvements.

> For more information, see: David A. Garvin, *Managing Quality* (New York: The Free Press, 1988); Kioshi Suzaki, *The New Shop Floor Management* (New York: The Free Press, 1993); Armand V. Feigenbaum, *Total Quality Control*, Third Edition, Revised (New York: McGraw-Hill, 1991).

Coach This is any individual who serves as a guide or advisor, leading people through process improvement, usually through the application of quality management principles. The term also describes a manager who advises and supports employees instead of telling them what to do.

The idea of coaching suggests that managers guide and encourage their people, but they do not do their employees' jobs (just as football coaches cannot play the game for their players). When managers think of themselves as coaches, they are more likely to focus on what they can do to help employees better understand organizational work, how to function well together, and how to improve their personal and group abilities.

Members of process improvement teams are frequently not experts on the mechanics of improvement since the teams are made up of people who spend their time executing a process. Therefore, many organizations have found it practical to assign a trained coach (or "quality advisor") to help them in their process improvement work. The coach teaches the team the basics of data analysis, flowcharting, and design, and guides them through the cycle of process improvement or reengineering.

Managers must forget about controlling and directing; they must begin to teach and coach.

William Lareau

Such coaches understand the dynamics of teams and the basic tools of quality improvement.

> For more information, see: Jack D. Orsburn et al., *Self-Directed Work Teams: The New American Challenge* (Burr Ridge, IL: Irwin Professional Publishing, 1990); Thomas H. Berry, *Managing the Total Quality Transformation* (New York: McGraw-Hill, 1991); Max DePree, *Leadership Is an Art* and *Leadership Jazz* (New York: Doubleday Currency, 1989 and 1992); Peter R. Scholtes, *The Team Handbook* (Madison, WI: Joiner Associates, 1988); William Lareau, *American Samurai* (New York: Warner Books, 1991).

Coloration This refers to someone deliberately distorting information about a process. Often the cause of such distortions is someone using the information for self-serving purposes (for example, to make a process appear as performing better than the data indicates).

> For more information, see: Joseph M. Juran, *Juran on Planning for Quality* (New York: The Free Press, 1988).

Commitment This usually refers to an organization's or an individual's resolve to apply TQM-like values and practices. It often refers to organizations that have developed a strong strategic vision for themselves and are embarking on the application of fact-based quality management principles.

The committed person doesn't play by the "rules of the game." He is responsible for the game. If the rules of the game stand in the way of achieving the vision, he will find ways to change the rules. A group of people truly committed to a common vision is an awesome force. They can accomplish the seemingly impossible.

Peter M. Senge

Every book and presentation on quality-related topics discusses the importance of commitment to and faith in the principles of quality. This commitment is necessary because it will take time to change organizational culture and practices enough to see results from quality efforts. Always targeted for special attention is senior management who, if not committed, will, by their silence, guarantee minimal if no implementation of quality practices.

Senior management commitment is the most critical factor in the successful implementation of process improvement and TQM strategies because these managers create the organizational environment and culture to which all other employees adapt. If they say they want quality values and practices but are not committed to the implementation of these values and practices, employees will not be either.

We can contrast commitment with compliance. Commitment implies that a person internalizes and takes as his or her own a set of principles and values. Compliance suggests that an individual accepts certain principles and values to keep his or her job but feels no personal identification with these principles. TQM is much more likely to succeed in those organization that foster feelings of commitment by all employees.

> For more information, see: Rosabeth Moss Kanter, Barry A. Stein, and Todd D. Jick, *The Challenge of Organizational Change* (New York: The Free Press, 1992); Roger J. Howe, Dee Gaeddert, and Maynard A. Howe, *Quality on Trial*, Second Edition (New York: McGraw-Hill, 1995); Peter M. Senge, *The Fifth Discipline* (New York: Doubleday Currency, 1990).

Common Cause The phrase refers to one or more sources of variation in the output of a process that affect all individual results or values of output.

Common causes are always present in a process. In processes under statistical control, there will always be some variation, but this always falls within statistically predictable control limits and is inherent in the system. Reducing common cause variation requires modifying the steps and procedures in a process in one or more ways. In the practice of TQM, an appreciation of common cause variation is particularly important. In traditional management, variation due to common causes is often treated as a single event rather than the result of a pattern of activities among employees, machines, and the environment in which they operate. When this happens, managers address common cause variation as if due to special causes. In doing this, they are likely to add other elements to the process that can increase rather than reduce variation.

The notion of common versus special causes is central to any appreciation of W. Edwards Deming's views on management.

> In a common cause situation, there's no such thing as THE cause. It's just a bunch of little things that add up one way one day and another way the next. So we don't learn much by trying to find the differences between the high points and the low points when only common causes are at work.
>
> *Brian L. Joiner*

Without measuring, recording, and analyzing process data to understand the range of common cause variation in a process, any attempt at improvement will just be guesswork that is more likely to fail than succeed.

An example of common cause variation would be the diameter of a shaft from a machining process, which might vary from piece to piece by 3/1000 of an inch. By understanding the variation in the machining process, one can look at all the elements that contribute to that process to figure out how to reduce the variation. Another example of common cause variation would entail the time required for a word processing department to type letters. It may be that employees can expect to wait between three and four hours on a predictable basis. By examining the process that generates this kind of common cause variation, you can start to determine what changes in the process might improve this turnaround time. (See also *Chance Causes.*)

> For more information, see: W. Edwards Deming, *The New Economics* (Cambridge, MA: MIT Center for Advanced Engineering Study, 1993); Brian L. Joiner, *Fourth Generation Management* (New York: McGraw-Hill, 1994); Donald J. Wheeler, *Understanding Variation: The Key to Managing Chaos* (Knoxville, TN: SPC Press, 1993); William W. Scherkenbach, *Deming's Road to Continual Improvement* (Knoxville, TN: SPC Press, 1991).

Common Interest Group This is a network or group of people and/or organizations who have a common interest and share their knowledge. They may be part of a benchmarking consortium, members of a national association committee, or some industry group.

Communication This includes all the processes by which people in an organization exchange information. Open communication supported by different communication vehicles and technology is vital in implementing Total Quality Management. This is because the culture and environment of any organization are influenced greatly by formal and informal mechanisms of communication.

Communication is one of the processes that is most frequently mismanaged in organizations undergoing fundamental quality transformation. Although communication is typically treated as an after-thought, or as an ad hoc series of events, it is becoming clear that the same rigorous discipline applied to the improvement of any other process must also be applied to communication. Why? This is how employees find out what is going on and let others know what they are doing. It is an essential lubricant that makes the whole system work well. The lack of intelligent implementation of the communication process can doom a quality effort to mediocre results at best.

Without clear, complete, and honest communication and widely available channels for employees to gain information they need to play their roles in organizational processes, the system will be compromised. When people do not have accurate information, when managers keep information from employees, this results in employees making false assumptions about what is going on, and it foments mistakes. Information technology, including e-mail and voice-mail systems, is making company-wide access to information and communication much easier and is usually an important component in any company's implementation of TQM.

> For more information, see: Roger J. Howe, Dee Gaeddert, and Maynard A. Howe, *Quality on Trial*, Second Edition (New York: McGraw-Hill, 1995); Roger Tunks, *Fast Track to Quality* (New York: McGraw-Hill, 1992); William Lareau, *American Samurai* (New York: Warner Books, 1991); Frank K. Sonnenberg, *Managing with a Conscience* (New York: McGraw-Hill, 1994); James W. Cortada, *TQM for Information Systems Management* (New York: McGraw-Hill, 1995).

Company Wide Quality Control (CWQC) This means the application of quality practices in all processes within the enterprise. It involves all employees at all organizational levels. Often the term

> If you can't communicate with your employees, you won't be able to communicate with your customers. Clear, consistent communication is a key to developing employees who are able to meet or exceed their customer's expectations.
>
> *Roger Tunks*

CWQC is used synonymously with Total Quality Management because the same or similar practices of quality improvement are at the heart of both terms.

The phrase originated in Japan, where it is used interchangeably with Total Quality Control. The word *control* in this phrase is used synonymously with the American use of the word *management*. In the United States, the phrase Total Quality Management (TQM) is more fashionable.

For more information, see: Masaaki Imai, *KAIZEN* (New York: McGraw-Hill, 1986).

Competition In its best sense, competition is the process by which individuals and organizations use the experience of each other to get better and better at what they do. In its worst sense, competition fosters an attitude of winning and losing. TQM teaches that competition is valuable because it keeps companies alert to their responsibility of better serving customers, and that by maintaining this focus, a company is most likely to succeed. Conversely, when companies focus on each other instead of customers, they are likely to behave in ways that undermine their long-term survival as they waste resources trying to destroy one another.

Three areas in which adversarial or competitive relationships are on the decline involve the move toward fewer suppliers for an organization, partnering (joint marketing schemes), and benchmarking. Manufacturers (as well as many other businesses) have found that by using fewer suppliers they can negotiate better prices, better delivery, and higher-quality components, thereby dramatically cutting costs due to waste, poor quality, and lost time. For example, allowing a supplier to see your production schedule and be responsible for ordering materials automatically when needed for production can greatly lower inventory costs for materials.

Partnering can allow two organizations that compete in a particular market to combine resources to be more cost effective and, from a marketing point of view, more successful, than as ri-

---

In TQC [or CWQC] the first and foremost concern is with the quality of people. Instilling quality into people has always been fundamental to TQC. A company able to build quality into people is already halfway toward producing quality products.

Masaaki Imai

Typically the management of a company spend a lot of time worrying about share of market. How big is our piece of the apple pie? How can we enlarge it at the expense of the competition? It would be better if all the competitors would use this time and energy to expand the market. They would all gain.

W. Edwards Deming

vals. This is a common strategy, for example, between U.S. and Japanese automobile manufacturers. Apple Computer and IBM have developed a partnering strategy in the development of PC architectures that are compatible with each other.

Benchmarking is a third opportunity where competition is partially set aside so that common processes are compared to each other. For example, it is not uncommon in the computer industry for manufacturing personnel to visit a rival's plant. It helps them become more efficient with processes in which both have something to learn. Such cooperation usually does not and should not compromise their competition in the marketplace.

> For more information, see: Gregory H. Watson, *Strategic Benchmarking* (New York: John Wiley & Sons, 1993); W. Edwards Deming, *The New Economics* (Cambridge, MA: MIT Center for Advanced Engineering Study, 1993); Peter M. Senge, *The Fifth Discipline* (New York: Doubleday Currency, 1990).

Competitive Analysis This refers to actions taken to understand the differences between products, programs, and performance among competitors. It is the study of the gap between your company's performance and that of your rivals.

In recent years, the benchmarking process has become an effective way to undertake competitive analysis. This had led to a new category of practices called competitive benchmarking. The goals of this process remain as they always have been—to understand who does what well and why, product by product, market by market, and to what degree of efficiency and success in the marketplace. (See other entries on *Benchmarking*.)

> For more information, see: Gregory H. Watson, *Strategic Benchmarking* (New York: John Wiley & Sons, 1993); Christopher E. Bogan and Michael J. English, *Benchmarking for Best Practices* (New York: McGraw-Hill, 1994); Michael J. Spendolini, *The Benchmarking Book* (New York: Amacom, 1992).

Completion Plan Ratio Often a project or process measurement, it indicates how well an organization is achieving certain milestones in a project or process. The use of a completion plan ratio is a useful approach for understanding how well projects are being managed.

For more information, see: Harold Kerzner, *Project Management: A Systems Approach to Planning, Scheduling, and Control,* Fourth Edition (New York: Van Nostrand Reinhold, 1992).

Complexity In a process, this refers to the extra steps, activities, or time that do not add value in its completion.

Complexity often refers to steps that provide for rework or downtime that takes up resources. The more complexity there is in a process, the more inefficient it is. Improvements and reengineering often focus on eliminating process complexity.

For example, consider a simple assembly process that includes bringing together five components to complete a prod-

FIGURE 19

All parts of the process below the dotted line are complexity. Adapted from Brian L. Joiner, *Fourth Generation Management* (New York: McGraw-Hill, 1994). Used with permission.

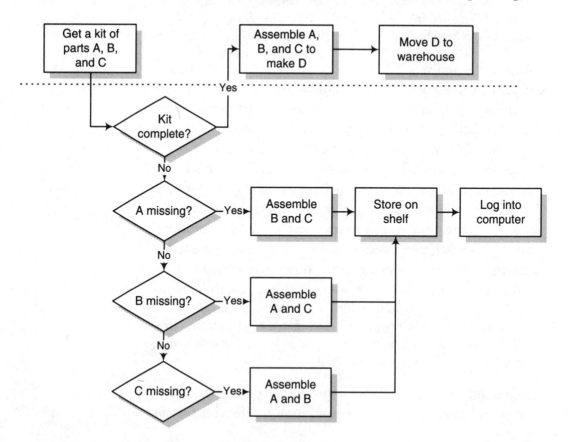

uct. If this process includes steps to take care of defective components, components missing at the time of assembly, or improperly assembled products, such steps create complexity and costs but do not add value. Process improvement would look at how to eliminate such problems and reduce the time taken to complete the assembly. Figure 19 illustrates a process that includes complexity and thus opportunities for improvement.

> For more information, see: Brian L. Joiner, *Fourth Generation Management* (New York: McGraw-Hill, 1994); W. Edwards Deming, *Out of the Crisis* (Cambridge, MA: MIT Center for Advanced Engineering Study, 1986).

Concurrent Engineering (CE) This is also called simultaneous engineering. It is a process for designing a product using all inputs and evaluations early on to make sure that all internal and external customers' needs and concerns are addressed.

Concurrent engineering allows for a faster product design cycle because it anticipates problems early on and balances the needs of all parties, including those of customers, engineering, suppliers, marketing, and manufacturing. The intent of this approach is to lead developers to weigh all elements of a product from the initial conception of an idea to the final disposition, taking into account such factors as quality, cost, and customer requirements, and expectations.

> For more information, see: H.G. Menon, *TQM in New Product Manufacturing* (New York: McGraw-Hill, 1992); William Lareau, *American Samurai* (New York: Warner Books, 1991); Joseph M. Juran and Frank M. Gryna, *Quality Planning and Analysis*, Third Edition (New York: McGraw-Hill, 1993).

Conformance This refers to a judgment that a product or service has conformed to a predetermined set of specifications, usually drawn from the requirements and desires of customers.

The idea originated in engineering and manufacturing, and is still used mainly in these two areas. In recent years, however, Philip Crosby and others have brought it into the main body of quality thinking. He defines quality simply as "conformance to requirements" of customers. This idea also applies to the work of employees who serve as suppliers to internal customers downstream from them. Their outputs should conform to the specifications and requirements of those who follow them in a process. This works best when specifications and requirements of each step in a process are properly documented and understood.

> For more information, see: Philip B. Crosby, *Quality Is Free* (New York: McGraw-Hill, 1979); Armand V. Feigenbaum, *Total Quality Control*, Third Edition, Revised (New York: McGraw-Hill, 1991).

Conforming Material This is material or parts that measurements show fall within specifications for quality.

Conformity This suggests that an output fulfills specifications, requirements, and quality standards. The term is generally used in manufacturing for raw materials and components as in the phrase, "The output is in conformity with specifications."

Conjoint Analysis This is a marketing research method that provides quantifiable assessments of the relative importance of one product, service, or feature in an offering over another. Sometimes people know this concept as tradeoff analysis.

The normal approach in this analysis is to ask customers to make choices between options in which one feature must be selected over another. For example, in a study of apartment rentals, a variety of features might include location, rent, and the availability of a swimming pool. Using conjoint analysis, researchers can quantify which of these is most important to come up with the best bundle of benefits for any price. This technique is very useful in circumstances where customer needs are in conflict and where a manufacturer, for example, needs to come up with a compromise solution that satisfies the greatest number of customers.

> For more information, see: Carl McDaniel, Jr. and Roger Gates, *Contemporary Market Research*, Second Edition (St. Paul, MN: West Publishing Company, 1993).

If you think about requirements, they are really only answers to questions. It all begins when you tell your people that you want them to "Do it right the first time" and they want to know what the "it" is.

Philip B. Crosby

Connector This is a line that links blocks in a flowchart to each other.

Conscious Errors This denotes any nonconformance to quality standards in a process that is the result of actions taken knowing they would cause errors. These are often defined as intentional human errors. Such errors can indicate problems in the culture of the organization that would foment such behavior.

Consensus This is a form of agreement reached by a group of people working on a plan or development of a process. It is *not* a majority vote to adopt a position. It also is not the boss's decision with everyone going along. Rather it results from discussion and other methods that allow for everyone's ideas to be considered and then bring the group to agreement on a particular decision or course of action they all can live with.

Consensus is particularly important in team behavior. Under this concept, members of the team agree to support and implement the decisions of the team as a whole, even if some do not agree fully with some of these decisions. There are a number of methods for reaching consensus, such as *nominal group technique* and *multivoting*. (See these terms).

> For more information, see: Peter R. Scholtes, *The Team Handbook* (Madison, WI: Joiner Associates, 1988); Jack D. Orsburn et al., *Self-Directed Work Teams: The New American Challenge* (Burr Ridge, IL: Irwin Professional Publishing, 1990); William Lareau, *American Samurai* (New York: Warner Books, 1991).

> Consensus is ... Finding a proposal acceptable enough that all members can support it; no member opposes it. It requires ... Time. Active participation of all group members. Skills in communication: listening, conflict resolution, discussion facilitation. Creative thinking and open-mindedness.
>
> *Peter R. Scholtes*

Conservatives This term identifies individuals who are reluctant or wary about applying quality principles and process management techniques based on the theory and logic of doing it.

They are often characterized as favoring "old thinking" instead of embracing "new thinking." People are often conservative because the environment in which they operate seems to them not to tolerate risk taking and trying new approaches. Or they may have become cynical by having gone through several different management fads. Conservatives become practitioners

American Samurai Insight: American management generally assumes that business practices and traditions are the way they are simply because that's the way it was meant to be; as if the current system is somehow mandated or preordained by the fundamental order of the universe. Nothing could be further from the truth.

William Lareau

of the new thinking when they more fully appreciate the insights of understanding organizations as systems and the techniques of TQM for managing systems more successfully.

For more information, see: Peter M. Senge, *The Fifth Discipline* (New York: Doubleday Currency, 1990); W. Edwards Deming, *The New Economics* (Cambridge, MA: MIT Center for Advanced Engineering Study, 1993); William Lareau, *American Samurai* (New York: Warner Books, 1991).

Constant Failure Rate Period This is that period during the life of a product or process when failures occur at a uniform rate. Engineers also call this the chance failure rate period. It is often characterized as the bottom of the *bathtub curve*. (See that term). It can only be established through careful measure of a process.

For more information, see: Joseph M. Juran and Frank M. Gryna, *Quality Planning and Analysis*, Third Edition (New York: McGraw-Hill, 1993); Armand V. Feigenbaum, *Total Quality Control*, Third Edition, Revised (New York: McGraw-Hill, 1991).

Sampling risks are of two kinds: (1) Good lots can be rejected (the producer's risk) and (2) Bad lots can be accepted (consumer's risk).

Joseph M. Juran and Frank M. Gryna

Consumer's Risk In any sampling plan, this is the chance that the customer will accept a lot of goods that does not meet specifications. The cost of that part of any lot not meeting requirements is a measurement of consumer's risk. Companies seek to use sampling plans that will minimize consumer's risk.

Engineers usually define consumer's risk in quantitative terms, often in a manufacturing environment, where they make purchases of large quantities of raw material or components on an ongoing basis. Sometimes people call consumer's risk "consumer's decision risk." (See also *Producer's Risk.*)

For more information, see: Joseph M. Juran and Frank M. Gryna, *Quality Planning and Analysis*, Third Edition (New York: McGraw-Hill, 1993); Armand V. Feigenbaum, *Total Quality Control*, Third Edition, Revised (New York: McGraw-Hill, 1991).

Continuous Improvement (CI) This is an important value statement that endorses the notion that we can continuously improve all processes and activities through the application of systematic techniques. It also embraces the idea that there should be a relentless, ongoing hunt to eliminate the sources of defects, inefficiencies, and nonconformance to customer specifications, needs, and expectations. The Japanese call CI *Kaizen*. Another definition calls CI a disciplined process that includes first, a commitment to excellence and second, carrying out the actual efforts to accomplish ongoing quality improvements in processes, services, and products.

An important foundation of continuous improvement is the notion of system variation and that, unchecked, variation will increase. This growing variation increases the chances that a process will deliver more defective outputs. Thus TQM emphasizes that managers must work at understanding system variation, its sources, and how to manage processes so as to reduce variation and improve quality.

Quality experts assert that opportunity for improvement is inherent in all processes, including those in control. Those processes considered in good condition (that are operating efficiently and with only common cause variation) usually require incremental improvement. The idea of continuous incremental improvement contrasts with reengineering, in which a current process is completely replaced with a radically new and improved one. Regardless of the approach to continuous improvement, it is important to appreciate that improvements can always be made and that doing so is vital to organizational survival.

Another foundation of CI is that whenever any product is launched, customers, competitors, and producer always learn two things about it: how well it works and how well it does not work. Any company that sits still and does not continuously improve the quality and utility of its offerings to customers will be a target for competing organizations that will be doing this. This may happen by adding benefits to the offering or even offering a whole new way of delivering those benefits. Or it may happen by improving processes and lowering the price, which on a per-dollar basis also enhances benefits.

Quality improvement requires you to continuously develop ways to do better work. Continuous improvement means "Fix it now!" and "Prevent problems before they happen" and "Look for new ways to meet customer needs." The theory of "If it ain't broke, don't fix it" has no place in a quality organization. Instead, the approach needs to be, "If it ain't broken, improve it."

V. Daniel Hunt

A standard way of considering the process of continuous improvement is the Plan-Do-Check-Act (PDCA) cycle, sometimes also called the PDSA cycle (S for *study* instead of check). This approach is designed to make everyone conscious of the importance of improvement and learning from experience. The PDCA cycle suggests that managers, teams, and all employees plan improvements in a process, implement them (sometimes in a pilot program), check how well they are working, make modifications if appropriate, and then implement them as standard policy for that process. PDCA is often represented as a wheel going up the hill of improvement. Figure 20 illustrates the PDCA wheel. (See also *Plan-Do-Check-Act* and *Kaizen.*)

FIGURE 20
The PDCA cycle.

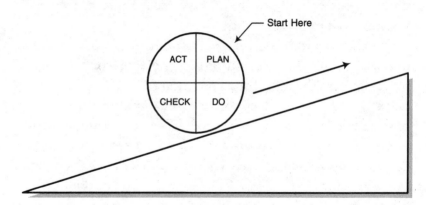

For more information, see: H. James Harrington, *Business Process Improvement* (New York: McGraw-Hill, 1991); Masaaki Imai, *KAIZEN* (New York: McGraw-Hill, 1986); Joseph M. Juran, *Juran on Leadership for Quality* (New York: The Free Press, 1989); Michael Hammer and James Champy, *Reengineering the Corporation* (New York: HarperBusiness, 1993); Peter Senge, *The Fifth Discipline* (New York: Doubleday Currency, 1990); Arthur R. Tenner and Irving J. DeToro, *Total Quality Management: Three Steps to Continuous Improvement* (Reading, MA: Addison-Wesley, 1992). (The idea of continuous improvement is covered in nearly every book on quality management. These books provide much how-to information.)

Continuous Sampling Plan This is an approach to sampling that is most appropriate for the output of processes that deliver a continuous flow of a product. In this plan, a company begins by inspecting 100 percent of the units coming from the process. After a certain number of items have been inspected with no defects, the plan goes to inspecting only a fraction of items. This continues until the company finds a nonconforming unit. At that point the plan reverts back to 100 percent inspection following the same pattern.

For more information, see: Armand V. Feigenbaum, *Total Quality Control*, Third Edition, Revised (New York: McGraw-Hill, 1991); Henry L. Lefevre, *Quality Service Pays: Six Keys to Success!* (Milwaukee, WI: ASQC Quality Press, 1989).

Control In quality management, this term refers to the idea that the outputs of a process fall within statistical control limits and that all variation is due to common causes.

We can contrast this meaning of control with traditional management, which focuses on control based on events and the behavior of individuals rather than on process performance through time. It does not take into account system and process capability. Trying to control people and make them meet a preconceived standard (not based on an understanding of system capability) is likely to introduce more uncertainty into organizational processes than improve them. It is also an approach more likely to inhibit improvement and innovation. It further encourages people to manipulate processes to achieve preconceived results.

Quality management, conversely, is less concerned with meeting some numerical standard than with understanding processes and how to make them work better. In considering the idea of control, it is important to understand that all processes are "in control." Even those processes that seem to operate very inefficiently, with a great deal of variation, are in control, given the capability of the system. This insight is why TQM focuses on *improving processes* rather than looking to bring individual elements into conformance with some arbitrary numerical standard that may have nothing to do with process capability or improvement.

> What part of the process requires continuous control? What part requires control only at specific stages? Where is preventive control needed, or at least control at a very early stage? And where is control essentially remedial? These questions are rarely asked when designing a control system. Yet, unless they are being asked—and answered—a control system that truly satisfies the needs of the work process cannot be designed.
>
> *Peter F. Drucker*

For more information, see: W. Edwards Deming, *Out of the Crisis* (Cambridge, MA: MIT Center for Advanced Engineering Study, 1986); Armand V. Feigenbaum, *Total Quality Control*, Third Edition, Revised (New York: McGraw-Hill, 1991); Kaoru Ishikawa, *Introduction to Quality Control* (Tokyo, Japan: 3A Corporation, 1990); Brian L. Joiner, *Fourth Generation Management* (New York: McGraw-Hill, 1994); Peter F. Drucker, *Management: Tasks, Responsibilities, Practices* (New York: Harper & Row, 1973).

Control Chart This is a graph and a statistical process control method used to track the capability and performance of a process over time.

> The control chart sends statistical signals, which detect existence of a special cause (usually specific to some worker or group or to some special fleeting circumstance), or tell us that the observed variation should be ascribed to common causes, chance variation, attributable to the system.
>
> *W. Edwards Deming*

Using a control chart, managers graph measurements or counts of the outputs from any kind of process. Then, using some simple statistical calculations, they can calculate the upper and lower control limits for the process (usually placed 3 standard deviations above and below the average or center line). A control chart is a way to check variation in a process and to discover the range of variation in the system when it is under control. Understanding this range of variation provides teams with a good sense of the process's capabilities. It also helps managers identify special causes of variation and distinguish these from common causes. Deming points that just because a process is in control does not mean that it can deliver outputs that meet specifications.

Control charts are one of the most widely used tools in quality management because they help people identify how well a process is working and how it might be improved. These charts can be used in every facet of an organization's operations and as the major reporting tool for results of processes. It is not uncommon, for example, to see secretaries, receptionists, grounds keepers, or police officers monitoring their activities with control charts. Long before the quality movement, people were using control charts on a selective basis, mostly in manufacturing. They have been extended to all types of activities that organizations now recognize and manage as processes. Figure 21 shows a simple control chart, with process measurements and upper and lower control limits and center line.

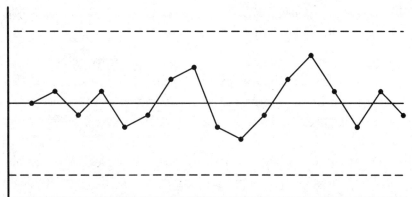

FIGURE 21
A typical control chart.

For more information, see: W. Edwards Deming, *Out of the Crisis* (Cambridge, MA: MIT Center for Advanced Engineering Study, 1986); Brian L. Joiner, *Fourth Generation Management* (New York: McGraw-Hill, 1994); William W. Scherkenbach, *Deming's Road to Continual Improvement* (Knoxville, TN: SPC Press, 1991); Henry L. Lefevre, *Quality Service Pays: Six Keys to Success!* (Milwaukee, WI: ASQC Quality Press, 1989).

Control Chart Factor This is a factor based on mathematical probability that facilitates the calculation of control limits on X-bar and range charts.

For more information, see: Hy Pitt, *SPC for the Rest of Us* (Reading, MA: Addison-Wesley, 1994).

Controllability This always denotes the extent to which any process meets criteria for self-control. Controllability is important and is related to statistical process control and to empowerment.

It implies first that management make sure that workers have the knowledge and tools they need to maintain a process in statistical control and that they can deal with nonconformances. It then suggests that workers have the authority and responsibility to manage the processes with which they are charged. In that case, we can say that a process has controllability built into it. A problem in many companies is that while managers have not built controllability into their processes, they still blame workers when something goes wrong, as if these workers had control.

For more information, see: Joseph M. Juran, *Juran on Leadership for Quality* (New York: The Free Press, 1989).

Control Limit This is a set of lines on a control chart equal distant above and below the center line. The position of the control limits is calculated by measuring outputs over time and shows the expected amount of variation we can expect in a process.

Control limits help us understand what a process is capable of. They show whether a process, given its current capability, will have a high or low level of variation in outputs. We calculate control limits using real data from the process in action. The control limits are generally set at 3 standard deviations above and below the center line. Figure 21 (page 83) includes control limits as dotted lines above and below the center line. As long as outputs remain within the control limits, the process is in statistical control. All variation in such a process comes from common causes, inherent in the system. It is important to note that control limits are *not* specification limits. A process may not be able to deliver outputs that regularly fall within specification limits. Once we know process capability, as indicated by the position of the control limits, a standard improvement activity is to take actions that reduce the spread of variation.

> For more information, see: W. Edwards Deming, *Out of the Crisis* (Cambridge, MA: MIT Center for Advanced Engineering Study, 1986); Brian L. Joiner, *Fourth Generation Management* (New York: McGraw-Hill, 1994); and many other books that deal with statistical process control.

Control Station This is a quality-activity point in a process or physical inspection station that performs feedback activities within a process or procedure.

For example, a control station can be the last person on a production line documenting how many bottles of soft drinks come off the line properly filled. It can also be an inspection task done to ensure invoices are correct before employees release them for the computer to print out and mail.

Control limits are not based on what we would like a process to do. They are based on what the process is capable of doing. They are computed from data using statistical formulas.

Brian L. Joiner

For more information, see: Joseph M. Juran, *Juran on Leadership for Quality* (New York: The Free Press, 1989).

Control Subject This is anything to which management attends to maintain or improve quality. Control subjects usually include process steps or product and service features. Joseph M. Juran adds that it is any feature "for which there is a quality goal." Thus, in editing this book to be a high quality publication, it is our control subject.

For more information, see: Joseph M. Juran, *Juran on Leadership for Quality* (New York: The Free Press, 1989).

Conventional Cost System ABC accounting approaches typically use this phrase to describe traditional cost accounting systems. Pre-ABC accounting systems relied on direct material and labor consumed as the way to apportion overhead. In ABC accounting practices, costs are apportioned to activities.

For more information, see: Peter B.B. Turney, *Common Cents: The ABC Performance Breakthrough* (Portland, OR: Cost Technology, 1991); Michael C. O'Guin, *The Complete Guide to Activity-Based Costing* (Englewood Cliffs, NJ: Prentice Hall, 1991).

Coonley-Agnew Process This is a process used to resolve differences among people in which they must (1) identify the area of agreement and disagreement, (2) come to a consensus on why they disagree, and (3) decide what actions to take concerning their disagreements. It was developed by Howard Coonley and P.G. Agnew and presented in a paper published in 1941.

Variations of these steps make up most contemporary negotiating strategies, such as the widely used Harvard "Getting to Yes" approach. Process and project teams frequently engage in similar discussions as they work through various issues, improvement strategies, and interpretation of data.

For more information, see: Roger Fisher and William Ury, *Getting to Yes* (New York: Viking Penguin, 1991); Roger Fisher and Scott Brown, *Getting Together* (New York: Viking Penguin, 1988); Joseph M. Juran, *Juran on Planning for Quality* (New York: The Free Press, 1988).

COPQ This is an acronym for *cost of poor quality*. (See this entry).

Core Competencies These are an organization's skills that are of strategic significance. They contribute to a competitive advantage over rivals. It is also a competence that is readily recognizable and accepted by customers.

For example, IBM has a core competence in the manufacture and marketing of computers, Wal-Mart in dry goods, and FedEx in overnight package delivery systems. Modern strategic planning calls for a careful understanding of core competencies. Benchmarking has proved to be a useful tool for a company to measure its core competencies and to understand how to improve.

There has been a debate over the past two decades about the appropriateness of companies venturing beyond their core competencies in their quest for growth and profits. This debate has not been resolved. However, the most important issue for growth and profitability in any company is the match between the ability to operate efficiently and effectively and the needs of some segment of customers. The successful companies leverage that ability to deliver more products and services that satisfy customers.

For more information, see: James Brian Quinn, *Intelligent Enterprise* (New York: The Free Press, 1992); Tom Peters, *Liberation Management* (New York: Alfred A. Knopf, 1992).

Core Ideology This incorporates the primary beliefs of an organization. It is a statement that says, "This is what we stand for."

In the great and enduring companies, this ideology always transcends the standard purpose "to make money and increase shareholder wealth." Such companies nearly always have ideologies that identify the role they believe they play in society, and what contribution they think they make. For example, for Boeing the core ideology is about pioneering aviation. This provides direction for the company, sometimes when standard P&L analysis would suggest otherwise, as was the case in building the

747 jet. Quality management principles are frequently standard practice in these companies.

> For more information, see: James C. Collins and Jerry I. Porras, *Built to Last* (New York: HarperBusiness, 1994).

Core Values This is a small set of enduring tenets that serve as general guiding principles for an organization. They represent fundamental assumptions about what is truth as far as the company is concerned. These values shape attitudes, culture, and behaviors in all companies, for better or worse.

In the great companies, core values are well articulated and honored. They are not subject to compromise for short-term financial gain or expediency. For example, the core values of GE as reported by Collins and Porras (see reference that follows) are the following:

- Improving the quality of life through technology and innovation.
- Interdependent balance between responsibility to customers, employees, society, and shareholders (no clear hierarchy).
- Individual responsibility and opportunity.
- Honesty and integrity.

Other companies have their unique values. What they all share is an uncompromising commitment to their values and to their employees and customers.

> For more information, see: James C. Collins and Jerry I. Porras, *Built to Last* (New York: HarperBusiness, 1994); Peter M. Senge et al., *The Fifth Discipline Fieldbook* (New York: Doubleday Currency, 1994).

In a visionary company, the core values need no rational or external justification. Nor do they sway with the trends and fads of the day. Nor even do they shift in response to a change in marketing conditions.

James C. Collins and Jerry I. Porras

Correctable Cause This is any tangible source of nonconformance in the outputs of a process that we can identify and eliminate.

Well-defined and measured processes frequently will generate nonconforming outputs that have correctable causes. This is because measures will signal *when* and *to what extent* things are different. Therefore, a process owner can identify the source of a problem, often using a root cause analysis technique, which leads to a solution. Correcting a cause helps return the process to a stable condition, operating as expected within controllable limits.

> For more information, see: W. Edwards Deming, *Out of the Crisis* (Cambridge, MA: MIT Center for Advanced Engineering Study, 1986); Brian L. Joiner, *Fourth Generation Management* (New York: McGraw-Hill, 1994).

Correction This term refers to all actions taken to reduce or eliminate causes of variations in process outputs.

> For more information, see: Henry L. Lefevre, *Quality Service Pays: Six Keys to Success!* (Milwaukee, WI: ASQC Quality Press, 1989).

Corrective Action This refers to actions taken to eliminate problems that adversely affect quality in products, services, and processes. These are also actions taken to bring a process back into conformance with requirements or to perform within a range of preset specifications. Figure 22 illustrates an ongoing loop for taking corrective actions in any process.

FIGURE 22

Corrective action loop. Adapted from John L. Hradesky, *Productivity and Quality Improvement* (New York: McGraw-Hill, 1988), p. 120.

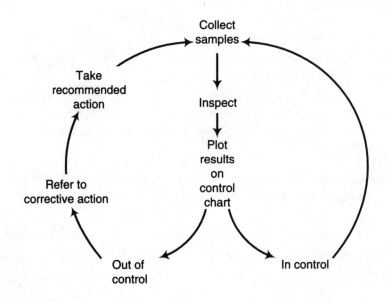

For more information, see: John L. Hradesky, *Productivity and Quality Improvement* (New York: McGraw-Hill, 1988); Joseph M. Juran, *Juran on Leadership for Quality* (New York: The Free Press, 1989).

Cost Analysis This is the act of simulating the costs of multiple options, such as various steps that could be taken to improve or reengineer a process. Process improvement work often requires that we review and understand what various alternatives might cost and yield. Besides helping to justify an option, this analysis helps managers set goals and expectations for costs and can become input into accounting and budgeting activities.

For more information, see: Peter B.B. Turney, *Common Cents: The ABC Performance Breakthrough* (Portland, OR: Cost Technology, 1991).

Cost Assignment View This view or perspective on ABC accounting practices is about assigning costs to specific activities and then assigning specific activities to cost objects.

The purpose of this view is to offer data about activities, resources, and cost objects (the drivers of costs). With contemporary ABC methods, it is possible to identify the costs of all activities, to assign dollar amounts to all resources associated with those activities, and then to link both to cost objects.

For more information, see: Peter B.B. Turney, *Common Cents: The ABC Performance Breakthrough* (Portland, OR: Cost Technology, 1991).

Cost Driver This is any factor that affects or changes the cost of any activity or process. When people change how they execute an activity, this alters the amount of resources required to perform the activity or process and thus the costs associated with this activity.

For instance, supplies used in an activity influence the amount of other resources (for example, machines, people) needed to perform that activity. Activities or processes often have more than one cost driver. The concept of a cost driver is an important one in the implementation of ABC accounting systems. This is because we see that cost drivers are the source of expenses as opposed to the products or people used in more conventional accounting approaches. Therefore, the identification and measurement of cost drivers are important activities in both financial and operational terms.

> For more information, see: Peter B.B. Turney, *Common Cents: The ABC Performance Breakthrough* (Portland, OR: Cost Technology, 1991).

Cost Element This is always designated as the amount of money paid for a resource, whether this is a person, product, or service. In ABC accounting, cost elements are assigned to a process or activity. They are also considered part of an activity cost pool. Cost elements are basic building blocks in the world of ABC accounting.

> For more information, see: Peter B.B. Turney, *Common Cents: The ABC Performance Breakthrough* (Portland, OR: Cost Technology, 1991).

Cost Object This is a term from ABC accounting that denotes the reason or cause for the performance of any activity. A customer can be a cost object (customers are an important reason for activities in an organization; thus they are cost objects), so can products, services, activities, and obligations. Each of these is a reason why an organization might expend money. For example, an employee performs a particular activity (such as makes a sales call—and thus expends resources) because a customer wants to see him. This customer becomes the cost object.

ABC accounting always assigns costs to cost objects. Activity drivers form the basis for expenses charged to particular cost objects. ABC seeks to measure as accurately as possible the amount of activity the organization "consumes" to satisfy a cost object. With such an approach you can assign costs and measure performance. From a quality management viewpoint, a cost object provides the motivation for all process activity. Knowing what that activity will cost given the object, allows you to decide, in relation to alternatives, if it is worth doing. This approach measures how valuable cost objects are to an enterprise. Finally, this approach helps quantify the costs and opportunities for improvement within a process by giving managers a better sense of whether various activities will yield a value for or in the cost object that will justify the investment.

> For more information, see: Peter B.B. Turney, *Common Cents: The ABC Performance Breakthrough* (Portland, OR: Cost Technology, 1991); Michael C. O'Guin, *The Complete Guide to Activity-Based Costing* (Englewood Cliffs, NJ: Prentice Hall, 1991).

Cost Object Activity This is any activity that enhances cost object value. It is another way of saying any action that is of benefit to a customer or enhances the attractiveness and value of any product or process.

> For more information, see: Peter B.B. Turney, *Common Cents: The ABC Performance Breakthrough* (Portland, OR: Cost Technology, 1991).

Cost of Poor Quality (COPQ) These are the extra expenses caused by delivering poor quality goods or services to customers. Said another way, these are the costs involved in the delivery of outputs to customers that would disappear if there were no quality problems.

These expenses have two sources: (1) internal failure costs (from defects before customers get the product) and (2) external failure costs (costs after a customer receives the poor product or service). Rework, repairs, lost future business, and warranty payments are all examples of costs associated with poor quality.

The cost of poor quality is often very large, and most managers do not understand it well. Various studies of the cost of poor quality show that it is often 20 to 25 percent of a company's revenues and that even well-run, quality-focused enterprises have costs of bad quality that are over 10 percent. Conventional wisdom holds that these companies can reduce these costs by roughly 40 percent over five years with a focused quality program. Examples of the cost of poor quality include (1) lost business because of an irate customer who never buys from you again, (2) a faulty $35 part in an automobile that causes the re-call of 700,000 vehicles for what is now a $130 repair (due to cost of labor and a new part), or (3) any refund or credit. Another example is the salary paid an employee for doing things more than once, such as retyping a letter. The costs of poor quality are like a iceberg: A few are apparent and direct, and many more lie hidden and indirect. Figure 23 captures this idea.

FIGURE 23

The direct and indirect costs of poor quality. Adapted from J. M. Juran and Frank M. Gryna, *Quality Planning and Analysis*, Third Edition (New York: McGraw-Hill, 1993) p. 23.

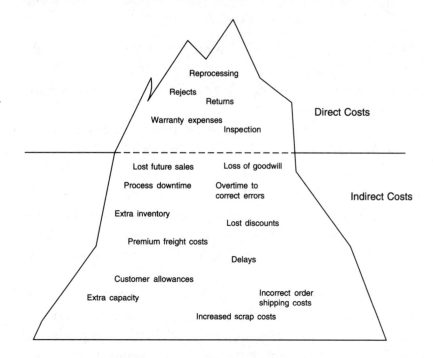

For more information, see: David Garvin, *Managing Quality* (New York: The Free Press, 1988); Thomas H. Berry, *Managing the Total Quality Transformation* (New York: McGraw-Hill, 1991); Philip B. Crosby, *Quality Is Free* (New York: McGraw-Hill, 1979); Joseph M. Juran and Frank M. Gryna, *Quality Planning and Analysis*, Third Edition (New York: McGraw-Hill, 1993); Joseph M. Juran, *Juran on Leadership for Quality* (New York: The Free Press, 1989); William Lareau, *American Samurai* (New York: Warner Books, 1991).

Cost of Quality (COQ) This refers to all costs involved in the prevention of defects, assessments of process performance, and measurement of financial consequences. Management uses this to document variations against expectations and as a measure of efficiency and productivity. In short, it is the cost justification of quality efforts.

The cost of quality represents positive activities in an organization. This is because we can view investing in these activities as what is necessary to deliver the quality customers require at the lowest cost possible. Making intelligent investments in quality helps generate additional business for an organization while reducing the costs associated with poor quality. For example, higher-quality products might allow an organization to enhance the value of its offering to customers with a no-questions-asked money-back guarantee. Why? Because the quality is so high that it will expect few of those customers to have to exercise the guarantee. Philip Crosby's approach to quality management is based on the idea that quality is free. By this he means that by attending to all the activities that cause quality deviations, the company will actually lower costs because it is always cheaper to do it right the first time than it is to repair something after the fact. Figure 24 illustrates the difference between a traditional view of the costs of quality before and after taking a TQM approach.

> The cost of quality takes the business of quality out of the abstract and brings it sharply into focus as cold hard cash. Suddenly the potential for achievement is there. Suddenly it really is a profit maker instead of a negative thought.
>
> *Philip B. Crosby*

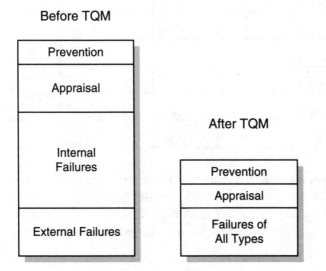

Before TQM

| Prevention |
| Appraisal |
| Internal Failures |
| External Failures |

After TQM

| Prevention |
| Appraisal |
| Failures of All Types |

FIGURE 24

The relative costs of quality before and after TQM.

For more information, see: Philip Crosby, *Quality is Free* (New York: McGraw-Hill, 1979); Armand V. Feigenbaum, *Total Quality Control*, Third Edition, Revised (New York: McGraw-Hill, 1991); William Lareau, *American Samurai* (New York: Warner Books, 1991); V. Daniel Hunt, *Quality in America* (Burr Ridge, IL: Irwin Professional Publishing, 1992).

Count Chart We use this type of control chart to evaluate the stability of a process by counting events of a particular type that occur or that you can observe in a sample. This is often called a "c chart."

This is a simple and common tool. Its fundamental purpose is as a data collection device for events that can be observed and counted. After a person has collected data concerning several samples, this can serve as the input for the development of control charts. Figure 25 shows a simple count of the number of times pizza was delivered to customers within 30 minutes of the order being placed, over one week.

FIGURE 25

Count chart for pizzas delivered within 30 minutes.

	Pizzas Delivered	W/In 30 Minutes	Percent
Day 1	98	83	84.7%
Day 2	115	102	88.7%
Day 3	75	69	92.0%
Day 4	125	119	95.2%
Day 5	87	78	89.7%
Day 6	103	96	93.2%
Day 7	145	132	91.0%

For more information, see: Howard Gitlow et al., *Tools and Methods for the Improvement of Quality* (Burr Ridge, IL: Irwin, 1989); Donald J. Wheeler and David S. Chambers, *Understanding Statistical Process Control*, Second Edition (Knoxville, TN: SPC Press,

1992); Eugene L. Grant and Richard S. Leavenworth, *Statistical Quality Control*, Sixth Edition (New York: McGraw-Hill, 1988).

Countdown This is the act of performing a series of events or steps in a predetermined sequence. The countdown is more than the simple act of counting backwards, for example, the number of seconds before a rocket takes off. It is the technique operators use to ensure that a correct sequence of events occurs before they let an anticipated event or result happen.

Managers can use the countdown when involved in structured processes that require us to take each step in a certain order. It is often used where physical security or safety is an issue, but it is increasingly common in highly precise process steps in manufacturing and in the delivery of services.

Count-Per-Unit Chart This is a chart that registers the ratio of the number of events that occur to the total area of opportunity for them to occur. This is commonly called a "u chart." Whereas a count chart tracks the number of events or defects of a particular type in a sample, a u chart tracks defects of all types in a sample. A u chart is especially helpful if the areas for opportunity are of different sizes. It allows you to compare defects from lot to lot more accurately.

If you are inspecting paper as it is manufactured, for example, you might take a 200 square foot lot and inspect it for blemishes or defects of all types. If you consider a standard sample size to be 100 square feet, then a 200 square foot lot of paper represents two areas of opportunity. If you find eight blemishes in these two areas of opportunity, your ratio is 8/2, or 4 blemishes per 100 square feet. From the collection of such data, operators can develop control charts that record these ratios for each area of opportunity to understand process capability and identify opportunities for improvement.

For more information, see: Howard Gitlow, et al., *Tools and Methods for the Improvement of Quality* (Burr Ridge, IL: Irwin, 1989);

Donald J. Wheeler and David S. Chambers, *Understanding Statistical Process Control*, Second Edition (Knoxville, TN: SPC Press, 1992); Eugene L. Grant and Richard S. Leavenworth, *Statistical Quality Control*, Sixth Edition (New York: McGraw-Hill, 1988).

Cp This signifies process capability index. See *Capability Process Index (Cp)* for explanation.

For more information, see: George L. Miller and LaRue L. Krumm, *The Whats, Whys & Hows of Quality Improvement* (Milwaukee, WI: ASQC Press, 1992); John L. Hradesky, *Productivity and Quality Management* (New York: McGraw-Hill, 1988); Hy Pitt, *SPC for the Rest of Us* (Reading, MA: Addison-Wesley, 1994).

Cpk This signifies process capability index based on the actual rather than the potential of the process. See *capability process index (Cpk)* for explanation.

For more information, see: George L. Miller and LaRue L. Krumm, *The Whats, Whys & Hows of Quality Improvement* (Milwaukee, WI: ASQC Press, 1992); John L Hradesky, *Productivity and Quality Management* (New York: McGraw-Hill, 1988); Hy Pitt, *SPC for the Rest of Us* (Reading, MA: Addison-Wesley, 1994).

Criterion This is a standard (usually derived from factual information) on which to base a decision.

A criterion often includes standards for the needs of customers, standards for decreased cycle time and lower costs and waste. In well-managed processes, managers define different criteria in a thoughtful and formal manner since they influence the characteristics of any changes. Organizations usually determine what the various criteria should be based on research and analysis before they put them in place. Benchmarking is one activity they undertake to identify standards on which to base the criteria for any process.

For more information, see: Christopher E. Bogan and Michael J. English, *Benchmarking for Best Practices* (New York: McGraw-Hill, 1994).

Critical Defect Also known as a critical characteristic, this is a defect that, based on experience, is considered hazardous to employees or to have the potential to harm or injure end users of a product or process.

> For more information, see: Armand V. Feigenbaum, *Total Quality Control*, Third Edition, Revised (New York: McGraw-Hill, 1991).

Criticality Analysis This is a technique for managing the analysis of complex and varied data about the importance of a product or service feature. Managers might classify a feature, for example, as critical for a variety of simultaneous reasons, such as crucial to human safety, required by government regulations, financially attractive, or insisted upon by customers.

Another way to view criticality analysis is to see it as a process to make sure you identify product or service features that are vital to the ultimate success of the endeavor.

> For more information, see: Dev G. Raheja, *Assurance Technologies: Principles and Practices* (New York: McGraw-Hill, 1991).

Critical Path Process owners use this term to describe the shortest distance between the beginning and end of a process or activity. When managers document processes, figuring out the critical path helps them identify the important steps that must occur for the entire process to operate properly and efficiently. This term and idea are commonly used in project management.

The technique was first used to list the sequence of activities in a process that is most important for completing the entire process, and, most frequently, those steps upon which other steps depend. A critical path chart, often called a PERT (Program Evaluation and Review Technique) chart, documents the order and time required to complete all the main steps for executing a process to get a final output.

The critical path view of processes also has an additional purpose: to separate out those steps that employees can do concurrently from those that they must do sequentially so as to reduce the time it takes to perform a process. In manufacturing, for example, parallel activities might include building the engine for a car at the same time as the body frame. Looking at critical paths allows a process owner to ask tough questions about the possibility of eliminating certain steps to simplify the process, reduce costs, and decrease the time it takes to execute.

> For more information, see: H.G. Menon, *TQM in New Product Manufacturing* (New York: McGraw-Hill, 1992); George Stalk, Jr. and Thomas M. Hout, *Competing Against Time* (New York: The Free Press, 1990); Harold Kerzner, *Project Management: A Systems Approach to Planning, Scheduling, and Control,* Fourth Edition (New York: Van Nostrand Reinhold, 1992). Many organizations use the computer program, *Microsoft Project,* to analyze, document, and create a printed chart of the critical path for completing a project.

Critical Processes These are the processes deemed crucial to the success of an endeavor. These may be processes that, if not managed well, can create dangers to humans (for example, the handling of nuclear waste) or risk of serious financial troubles (for example, the marketing of an important product). They often refer to processes that are crucial to a business's success, such as new product development, order and delivery, and distribution. If these processes do not work well, it jeopardizes company survival.

> For more information, see: Joseph M. Juran, *Juran on Planning for Quality* (New York: The Free Press, 1988).

Critical Success Factors (CSFs) These are variables or activities that *absolutely* have to go right for a process, program, or a team to make its objectives. Traditionally a part of strategic planning, process owners frequently list CSFs during planning. They are the factors they must manage with extra care to ensure overall plan success.

In process reengineering projects, process owners spend a lot of time identifying CSFs. In these projects, owners design new process tasks and steps. Then they determine the factors on which the success of each step depends. Understanding strengths and weaknesses of a process helps ensure its successful execution. CSFs also lead to identification of issues outside a process that can influence its success or failure, such as special causes of variation in performance.

For more information, see: Thomas H. Berry, *Managing the Total Quality Revolution* (New York: McGraw-Hill, 1991).

Cross-Functional Team This refers to a process team composed of people from various departments or functional areas involved in executing a process. A cross-functional team works on solving a specific problem, developing a project or product, or improving an organizational process that requires the cooperation of various functional areas to execute. Such a team often has the mandate to develop a new process or to improve one already in place.

A good example of a cross-functional team at work is in product development. The team may consist of a product manager, a marketing manager, one or more design engineers, a manufacturing manager, and a sales manager. The purpose of bringing together employees from different functions to design a product is because all these areas have an interest in and can affect the success of the product in the market. Working together, the team can better plan a product that will meet customer needs, while also not incorporating features or designs that make it more costly or complicated to manufacture than is needed. Because most organizational work requires cooperation across the organization for improved efficiency and effectiveness, cross-functional teams are a good way to achieve this cooperation.

While the idea of cross-functional teams is sensible and works, its implementation can be problematic in company cultures where traditional functional departments with attendant functional department objectives remain in place. Such objectives are often at odds with the goals of a team requiring cooperation and communication across functions. Thus, for cross-functional teams to perform best, companies need to view and organize work around processes.

For more information, see: Jack D. Orsburn et al., *Self-Directed Work Teams: The New American Challenge* (Burr Ridge, IL: Irwin

Professional Publishing, 1990); Sarv Singh Soin, *Total Quality Control Essentials* (New York: McGraw-Hill, 1992); and Peter R. Scholtes, *The Team Handbook* (Madison, WI: Joiner Associates, 1988); Y.S. Chang, George Labovitz, and Victor Rosansky, *Making Quality Work* (New York: HarperBusiness, 1993).

Cross Plot This is a less common name for a scatter diagram or a scatter plot, used to study the relationship between two variables and to detect correlations.

Engineers often employ this tool. However, people from all areas of organizations are finding this to be a useful device for understanding data. For example, it can help in looking at information from two different sources, particularly if each source has information that can help predict the behavior of the other. In using this tool, it is important to note that while the diagram may show a relationship, it is not always evident that one factor *causes* the other. That requires further investigation and experimentation.

A simple example of the tool's use would be the customer wait time for a service representative in relation to the time of day. Figure 26 illustrates a cross plot for waiting time in minutes by time of day. This diagram illustrates a positive correlation between the time of day and waiting time. Understanding this correlation, the firm could put more representatives on the phones during the busiest times. Other examples of the use of a scatter diagram might be the relation between cutting speed and tool life or certain output defects in relation to assembly line speed or time on the shelf in relation to the efficacy of a medicine.

The relation of one factor with another might not always be predictable, and using cross plots or scatter diagrams helps an investigator determine if there is any kind of correlation between the two factors being reviewed. They may discover there is no correlation. However, if they do see a correlation, managers can use this information to learn how to make improvements by manipulating one or both of the factors (or other factors that affect these).

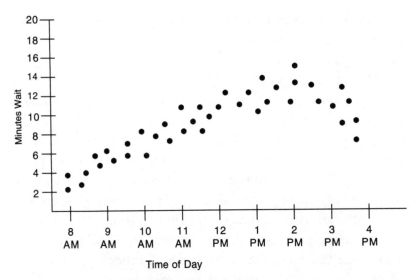

FIGURE 26

Cross plot scatter diagram showing correlation between minutes wait and time of day phone call is made.

For more information, see: Michael Brassard and Diane Ritter, *The Memory Jogger II* Methuen, MA: Goal/QPC, 1994); George L. Miller and LaRue L. Krumm, *The Whats Whys & Hows of Quality Improvement* (Milwaukee, WI: ASQC Quality Press, 1992); Hy Pitt, *SPC for the Rest of Us* (Reading, MA: Addison-Wesley, 1994).

Culture When used in the context of business, this defines the patterns of shared values, beliefs, and attitudes within an organization. These values, beliefs, and attitudes manifest themselves in the behaviors and actions of managers and employees. Thus, paying attention to and creating a culture that supports quality management is vital to transforming an organization and implementing quality management practices. Otherwise the use of TQM techniques will be mechanical and not likely to deliver the long-term results for customer satisfaction, more efficient processes, lower costs, growth, and profitability that it promises.

In enterprises introducing Total Quality Management practices, defining the differences between the existing and desired organizational culture becomes an important activity. One of the first steps in that process involves understanding existing values and attitudes of employees. From that baseline, managers can begin making changes in those values that encourage a strong commitment to customer satisfaction (and even delight)

Cultures can have powerful consequences, especially when they are strong. They can enable a group to take rapid and coordinated action against a competitor or for a customer. They can also lead intelligent people to walk, in concert, off a cliff.

John P. Kotter and James L. Heskett

and continuous improvement. Such a commitment then makes incorporating teamwork and process management a proactive part of everyone's job.

The successful implementation of TQM requires the creation of a culture, starting with top management, where values like the following are present: (1) an unrelenting focus on delivering quality to customers; (2) a focus on processes and their continuous improvement; (3) teamwork, cooperation, empowerment, and fairness among all departments and units of the company; (4) openness and sharing of information; (5) the use of scientifically derived data for making decisions.

Top managers must be involved because they are the ones who create the environment to which all other employees will adapt. Further, it is important to appreciate that employees never believe words unless they are accompanied by behaviors consistent with those words. Thus, top management must conscientiously "walk the talk" if they expect to transform their organization. The failure of "TQM programs" can inevitably be traced to cultures in which the values of TQM are not prevalent. TQM is not a program, it is an insight into what is a better way of managing the organization as a system to get results.

> For more information, see: Edgar H. Schein, *Organizational Culture and Leadership*, Second Edition (San Francisco: Jossey-Bass, 1992); Rosabeth Moss Kanter, Barry A. Stein, and Todd D. Jick, *The Challenge of Organizational Change* (New York: The Free Press, 1992); Marshall Sashkin and Kenneth J. Kiser, *Putting Total Quality Management to Work* (San Francisco: Barrett-Koehler Publishers, 1993); Peter M. Senge, *The Fifth Discipline* (New York: Doubleday Currency, 1990); Frank K. Sonnenberg, *Managing with a Conscience* (New York: McGraw-Hill, 1994); John O. Whitney, *The Trust Factor* (New York: McGraw-Hill, 1994); Craig R. Hickman, *Mind of a Manager, Soul of a Leader* (New York: John Wiley & Sons, 1990); John P. Kotter and James L. Heskett, *Corporate Culture and Performance* (New York: The Free Press, 1992); James C. Collins and Jerry I. Porras, *Built to Last* (New York: HarperBusiness, 1994); James Champy, *Reengineering Management* (New York: HarperBusiness, 1995).

Cultural Needs This refers to employees' needs for job security, self-respect, the respect of others (recognition), continuity of patterns of behavior of the enterprise, and shared cultural values.

These are important elements that must be considered during any organization's effort to change its culture. Each requires its own set of processes and action plans to implement; each can be managed using the same disciplines employed in process improvement.

> For more information, see: Thomas H. Berry, *Managing the Total Quality Transformation* (New York: McGraw-Hill, 1991); Rosabeth Moss Kanter, Barry A. Stein, and Todd D. Jick, *The Challenge of Organizational Change* (New York: The Free Press, 1992); Marshall Sashkin and Kenneth J. Kiser, *Putting Total Quality Management to Work* (San Francisco: Barrett-Koehler Publishers, 1993); John P. Kotter and James L. Heskett, *Corporate Culture and Performance* (New York: The Free Press, 1992).

Cultural Pattern This is any collection of beliefs, habits, practices, and experiences that define the activities and attitudes of an organization. By careful examination, you can begin to discern such patterns and determine their contribution or deterrence in developing a culture for the implementation of TQM.

An especially useful tool for doing this is the feedback loop, which allows you to see the influential relationships among people, behaviors, expectations, and processes, and whether they reinforce or limit a company's achievement of its goals.

> For more information, see: Peter M. Senge, *The Fifth Discipline* (New York: Doubleday Currency, 1990); Peter M. Senge et al., *The Fifth Discipline Fieldbook* (New York: Doubleday Currency, 1994); John P. Kotter and James L. Heskett, *Corporate Culture and Performance* (New York: The Free Press, 1992).

Cultural Resistance This is any collective resistance to change caused by a conflict between the proposed change and existing organizational beliefs and practices.

Cultural resistance has become a problem as corporations and government agencies transform themselves into smaller enterprises that deploy process management techniques and become more focused on customer satisfaction. The best way to deal with cultural resistance is to acknowledge its source— people wanting to avoid uncertainties—and then keep employees deeply involved in proposing as well as implementing cultural change. In this way, employees and managers are most likely to embrace the cultural changes suggested by TQM.

> For more information, see: Warren H. Schmidt and Jerome P. Finnigan, *TQManager* (San Francisco: Jossey-Bass Publishers, 1993); Peter M. Senge, *The Fifth Discipline* (New York: Doubleday Currency, 1990); Peter M. Senge et al., *The Fifth Discipline Fieldbook* (New York: Doubleday Currency, 1994); John P. Kotter and James L. Heskett, *Corporate Culture and Performance* (New York: The Free Press, 1992).

Cumulative Sum Control Chart This type of chart (frequently called a CuSum chart) plots a value that is the cumulative sum of deviations of successive samples taken from some target value. The ordinate of each point represents the sum of the previous ordinates plus most recent deviation from the target specification.

The CuSum chart helps to detect slight but sustained shifts of the process average away from a target value or specification and helps the process owner become aware of problems that might be in a process. This means the owner can make changes sooner rather than wait until more substantial variation has been detected. To use a CuSum chart, you need to include what is a called a V-mask, which, when laid over the plotted points, helps the analyst identify if some points are not in statistical control. The diagonal lines on this mask indicate the upper and lower control limits of the process. You place the mask so that the inside center line is lined up with the last point on the chart. For example, if the width of rolled steel is to be .1 inch, we can use a cumulative sum control chart to detect any sustained deviations from this width. In figure 27, we see a sustained shift to the upper right of the chart, and using the V-mask, we note that point 8 falls outside the control limits indicated by the mask. This indicates the process may not be in control.

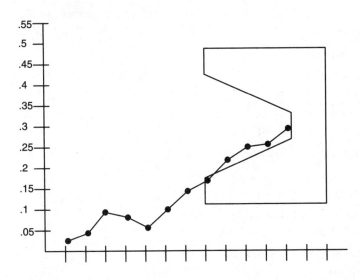

Width	.13	.11	.15	.09	.08	.15	.14	.13	.15	.12	.11	.10
Deviation	.03	.01	.05	.01	.02	.05	.04	.03	.05	.02	.01	0
Cum Sum	.03	.04	.09	.08	.06	.11	.15	.18	.23	.25	.26	.26

FIGURE 27

Cumulative sum chart. Note that at point 8 the mask covers the measurement, indicating that at that point the process was not in statistical control.

The value of CuSum is that while an entire process might look to be in control if one were to use a standard control chart, the CuSum chart provides a finer view of process outputs. The principle of cumulative sum control applies to any process in which one needs to know as early as possible in the execution of the process whether there is some slight variation from what is expected that might help to identify and correct problems.

For more information, see: N. Logothetis, *Managing For Total Quality: From Deming to Taguchi and SPC* (Englewood Cliffs, NJ: Prentice Hall, 1992); Eugene L. Grant and Richard S. Leavenworth, *Statistical Quality Control*, Sixth Edition (New York: McGraw-Hill, 1988); Hy Pitt, *SPC for the Rest of Us* (Reading, MA: Addison-Wesley, 1994).

Cumulative Uptime and Yield This is a way of analyzing and measuring the productivity of a process for reducing cycle time (the amount of time needed to deliver a finished output).

George Stalk and Thomas Hout (see reference that follows) describe it this way: "Each station in a sequential set of operations is either operating or not (that is, percent of uptime) and is turning out good product the first time or not (that is, percent yield). If every station had 100 percent uptime and 100 percent yield, the process would be running at its best possible cycle time. In most processes, there are some stations down, and others that are not turning out work up to specification. Consequently either the whole line shuts down or buffer inventories and queues are placed between stations. Either way, process cycle time is lengthened." (p. 207)

Understanding cumulative uptime and yield in a process gives a baseline for reducing cycle time. Even a 1 percent increase in cumulative uptime and yield at each station will be compounded throughout the entire process such that in a process with 19 steps, for example, this increase would raise the capacity of the entire process more that 35 percent.

> For more information, see: George Stalk, Jr. and Thomas M. Hout, *Competing Against Time* (New York: The Free Press, 1990).

Customer In quality management, this is anyone who receives the benefits of the output from another individual or organization. This definition includes both external (Big C) customers and internal (little c) customers. In both cases, delivering what customers value and need gives purpose and direction to productive activities.

One useful way to understand customer is to look at the root of this word, which comes from the same root as "custom." This implies that to meet needs, one must understand the importance of *customizing* outputs to the needs and expectations of *customers*. Making customer focus, of both external customers and internal customers, the defining aspect of all jobs can play an important role in shaping and changing organizational culture. It causes people to move from being inwardly directed and pleasing the boss to being externally directed and pleasing customers. It helps to instill a value of cooperation and service, both of which are consistent with making operations work more efficiently and improve, and to delivering better products and services to those who supply the funds for an organization to prosper—customers. Even in government this perspective is taking hold. The consequence of this is that government employees focus on providing services to taxpayers instead of following bureaucratic, self-serving rules.

> For more information, see: Ron Zemke and Dick Schaaf, *The Service Edge* (New York: New American Library, 1989); Christian Gronroos, *Service Management and Marketing: Managing the Moment of Truth in Service Competition* (Lexington, MA: Lexington Books, 1990); Valarie A. Zeithaml, A. Parasuraman, and Leonard L. Berry, *Delivering Quality Service* (New York: The Free Press, 1990); Carl Sewell, *Customers For Life: How to Turn That One-Time Buyer into a Lifetime Customer* (New York: Doubleday, 1990); David Osborne and Ted Gaebler, *Reinventing Government: How the Entrepreneurial Spirit Is Transforming the Public Sector* (Reading, MA: Addison-Wesley, 1992); Richard C. Whiteley, *The Customer-Driven Company* (Reading, MA: Addison-Wesley, 1991); Hal F. Rosenbluth and Diane McFerrin Peters, *The Customer Comes Second* (New York: William Morrow, 1992).

Customer Delight This phrase describes an emphasis on delivering products or services that exceed customers' requirements in ways they find valuable. It often helps us identify and create products and services customers may not have known they need but truly appreciate once they have them. Common examples from the recent past would be TV remote controls, microwave ovens, and many other devices and services that add value to a product or service that we did not anticipate we wanted before we had them, but that we come to expect afterwards.

Each of the major quality management contributors (Deming, Juran, Crosby, and several others) focuses on the need to direct all activities toward the satisfaction of customer requirements. Many organizations, however, have a conscious strategy for exceeding customer expectations, thus the term customer delight. Quality-disciplined enterprises will develop quantifiable goals for measuring customer delight. Doing this includes

> When you're close to your customers, you're on the way to a real competitive advantage. When their needs and expectations become the standards against which your organization measures its efforts and its heart, customers will find their expectations constantly exceeded. They'll experience delight. And they'll respond in a wonderful way—with loyalty.
>
> *Richard C. Whiteley*

creating rigorous customer satisfaction surveys, and processes that result in targeted customers being "delighted" with services and products. Delighting customers by exceeding their needs and expectations or providing products and services they did not know they needed is a long-term "constancy of purpose" strategy that helps keep a company profitable and growing.

In delighting customers, it is important to appreciate the four aspects of quality or value that organizations can focus on to deliver that quality. They are expressed in terms of utilities that a firm can deliver to customers: *form utility* (physical or tangible characteristics of the product or service that bring together its form and function); *time utility* (delivering offering when customers want it); *place utility* (delivering product where customers want it or making it convenient to get to a location where product or service is delivered); *possession utility* (includes price, terms, and warranties of satisfaction after purchase). By focusing on continuously improving and being innovative in these areas, organizations are more likely to take actions that will result in customer delight. Figure 28 captures these aspects of delighting customers.

FIGURE 28

Factors for delighting customers.

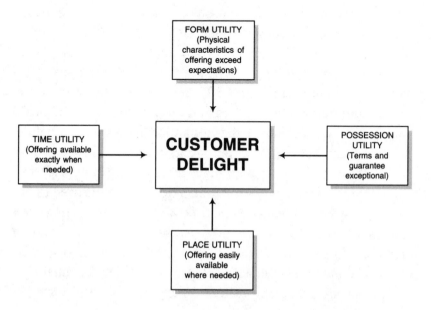

For more information, see: Brian L. Joiner, *Fourth Generation Management* (New York: McGraw-Hill, 1994); Tom Peters, *Liberation Management* (New York: Alfred A. Knopf, 1992); Y.S. Chang, George Labovitz, and Victor Rosansky, *Making Quality Work* (New York: HarperBusiness, 1993); Thomas H. Berry, *Managing the Total Quality Transformation* (New York: McGraw-Hill, 1991); Richard C. Whiteley, *The Customer-Driven Company* (Reading, MA: Addison-Wesley, 1991); Jonathan Barsky, *World-Class Customer Satisfaction* (Burr Ridge, IL: Irwin Professional Publishing, 1994).

Customer Requirements These are the benefits Big C and little c customers want from a product or service. They are a central aspect of quality: All processes and organizations should be thoroughly grounded in customer-defined requirements. Managers should develop processes to discover customer requirements in their products and services and make sure they incorporate those requirements into their offerings.

Related to this idea is the more forward looking approach of anticipating and helping to shape customer needs and working to delight them. In doing this, managers learn so much about their customers that they anticipate which additional features will be of benefit and will seek to effectively and efficiently deliver these benefits. This understanding and approach grow out of studies that show customer requirements for new products and additional levels of quality grow as the performance of a product or service improves.

Some companies seek to add benefits that exceed customer requirements, and this can be an important part of delighting customers. However, doing this sometimes causes companies to add features that add more cost to the product or service than customers will value. When this happens, the product does not meet customer requirements and is likely to fail. Meeting customer requirements thus becomes a careful balance of the benefits offered and costs.

For more information, see: Christian Gronroos, *Service Management and Marketing* (Lexington, MA: Lexington Books, 1990);

Valarie A. Zeithaml, A. Parasuraman and Leonard L. Berry, *Delivering Quality Service* (New York: The Free Press, 1990); Philip B. Crosby, *Quality Is Free* (New York: McGraw-Hill, 1979); Richard C. Whiteley, *The Customer-Driven Company* (Reading, MA: Addison-Wesley, 1991).

Customer Satisfaction This refers either to the delivery of a product or service that meets or exceeds customer expectations or requirements and to measurements that determine levels of satisfaction for the product or service provided.

For more information: see references listed under *Customer Delight.*

Customer/Supplier Analysis This is a collection of methods for providing insight into the needs and expectations of customers seen through the relationship between an organization and its suppliers. The goal of this analysis is to bring about the alignment of activities between suppliers and producers with customer requirements.

The analysis is a process; a commonly used worksheet provides information at a glance about what kind of alignment is necessary. The key to any successful supplier/customer analysis is listening to both, understanding the interrelationships of what is heard from both sides, and developing clear objectives of performance of a supplier based on that input. Figure 29 shows a simple form for undertaking this kind of analysis.

For more information, see: James H. Saylor, *TQM Field Manual* (New York: McGraw-Hill, 1992).

FIGURE 29

Worksheet showing the relationship of each supplier and customer in a process.

Input	Supplier	Requirement	MET/ NOT	Output	Customer	Expectation	MET/ NOT
Drawings	Engineering	Accurate and Complete		Documents	Carpenters	100% Accurate	

Customer-Supplier Partnership This is a term for describing a long-term relationship between a buyer and supplier in which mutual trust, teamwork, and dependence are typical.

In this kind of relationship vendors are considered an extension of a buyer's operations. Buyers use fewer suppliers and commit to long-term contracts and relationships. Suppliers meet quality standards of their customers. Quality managers employ this strategy to reduce operating and inspection costs, while improving quality of goods and services. This a well-proven and widely used strategy that has primarily been used by large manufacturing corporations. Thus, for example, an automotive manufacturer might only use three suppliers of tires whereas ten years ago it might have gone to fifteen suppliers and bought at the lowest price. Under the new arrangement, a buyer is dependent on a supplier to a greater extent and thus has a stake in making sure the supplier is well run and is financially successful.

Conversely, the supplier understands that its long-term business is dependent on it delivering high quality components to its main customers. Sometimes, functional boundaries may become blurred between the two organizations. For example, a tire supplier might have access to an automobile manufacturing plant's production schedule so it knows how many tires to have at that plant each day. That same supplier might also have access to the plant's checking account to deduct the cost of the tires sold each day.

> For more information, see: W. Edwards Deming, *Out of the Crisis* (Cambridge, MA: MIT Center for Advanced Engineering Study, 1986); George Stalk, Jr. and Thomas M. Hout, *Competing Against Time* (New York: The Free Press, 1990); Richard J. Schonberger, *Building a Chain of Customers* (New York: The Free Press, 1990); James F. Cali, *TQM for Purchasing Management* (New York: McGraw-Hill, 1993); William Lareau, *American Samurai* (New York: Warner Books, 1991).

Customer Value This is always defined as the difference between what a customer must pay for something and the value the

In supplier partnerships, contracts contain few penalties, incentives, and conditional clauses; there is a bond of trust between supplier and customer in which each assumes that the other will do his or her best.

William Lareauo

customer receives in return. Another way to state this is as "the difference between customer realization and sacrifice."

These terms—realization and sacrifice—can have a broad array of meanings from cost to quality, maintenance expense, or advantage gained. Strategies designed to satisfy customer requirements or to increase customer value focus on maximizing the difference between cost and value received. In process management, what customers value is frequently defined through surveys and other research tools and methods. An alternate way of understanding the idea of quality is the idea of *value*. It is a company's responsibility and purpose to continuously improve its ability to deliver value to customers and to make sure that what they deliver is what customers value. The components of value center around issues of form and function, time, place, price, warranties, and other aspects of service.

> For more information, see: Richard C. Whiteley, *The Customer-Driven Company* (Reading, MA: Addison-Wesley, 1991); Thomas F. Wallace, *Customer-Driven Strategy* (Essex Junction, VT: Oliver Wight Publications, Inc., 1992); Thomas H. Berry, *Managing the Total Quality Transformation* (New York: McGraw-Hill, 1991); William Lareau, *American Samurai* (New York: Warner Books, 1991).

Customer Value Analysis This is any analysis of customer activities designed to improve customer value while reducing the cost of delivering that value.

> For more information, see: Peter B.B. Turney, *Common Cents: The ABC Performance Breakthrough* (Portland, OR: Cost Technology, 1991).

Cycle Time This is the time it takes from beginning to end to complete a process. It is an important measure of process efficiency. TQM tools and techniques intelligently applied usually result in reduced cycle time.

Organizations recognize the value of time and that it is a limited resource. The strategic use of time is important in creating a competitive advantage. Part of the value customers pay for is having a product when and where they want it. In addition, the creation of products and services in less time lowers costs. Finally, long cycle times happen because of poor process planning and management. This exacerbates the up and down swings in customer demand that result from not having enough inventory of products when people want them and too much when they don't.

Cycle time reduction includes the careful analysis of steps in a process, noting how much time each step takes, and noting especially *queue times* and the efficiency of execution of each step. Eliminating steps that do not add value to the process and minimizing queue times by making improvements in the process are two ways to reduce cycle time. In the analysis of many processes, people find that the time taken to do work is a small fraction of the time taken to complete the process. The rest of the time comes from waiting in line to have that work done, either because of a pile-up from the previous step or because machines are down. Another source of time in a process is lead time (the amount of time required to resolve conflicts among outputs that require the same resources to produce). The study of cycle time reduction shows how to plan better and improve processes so as to eliminate bottlenecks and be more responsive to customers. (See also *Cumulative Uptime and Yield.*)

For more information, see: George Stalk, Jr. and Thomas M. Hout, *Competing Against Time* (New York: The Free Press, 1990); Gerard H. Gaynor, *Exploiting Cycle Time in Technology Management* (New York: McGraw-Hill, 1993).

> Time is the friend of the wonderful company, the enemy of the mediocre.
>
> *Warren Buffet*

Data These are facts presented in descriptive, numeric, or graphic forms. Organizations collect data either by taking measurements (called variable data) or making counts (called attribute data). Data become information when presented in some ordered format that makes these facts usable.

An important insight of TQM is that we should base decisions on facts rather than impressions or past experience. The concept of fact-based decision making is at the heart of efforts to improve or reengineer processes. About one-third of the effort expended on process improvement is typically associated with data gathering and analysis. (Statistical process control techniques provide the methods for doing this analysis.) The emphasis on fact-based decisions is having a profound effect on corporate cultures. It helps to eliminate arbitrary decisions and focuses employee attention on actions that will improve processes and the company's ability to better serve its customers.

For more information, see: W. Edwards Deming, *The New Economics* (Cambridge, MA: MIT Center for Advanced Engineering Study, 1993); Joseph M. Juran, *Juran on Leadership for Quality* (New York: The Free Press, 1989); John L. Hradesky, *Productivity and Quality Improvement* (New York: McGraw-Hill, 1988); Brian L. Joiner, *Fourth Generation Management* (New York: McGraw-Hill, 1994).

Data Bank This is any compilation of data organized to make it easy to retrieve either in paper files or on-line. In quality management, a data bank includes information about process improvement and "lessons learned" (Juran's term).

Gathering and storing information about how a process performs is critical to any attempt at improving performance. Such data banks include historical data, information on various types of control charts, and additional documentation on how a specific process works and has been changed and improved over time.

> For more information, see: John L. Hradesky, *Productivity and Quality Improvement* (New York: McGraw-Hill, 1988); Joseph M. Juran, *Juran on Planning for Quality* (New York: The Free Press, 1988).

Data/Statistical Analysis This analysis involves a collection of techniques and tools for gathering, sorting, organizing, and presenting data to create information that leads to understanding, sound decisions, and effective process management and improvement.

It is useful to appreciate that data only takes on meaning when it is ordered in particular formats or put "in formation." We can then use that information to understand situations and decide what to do. A problem of traditional management has been taking one or two pieces of data and using these to make decisions, as if they were information. TQM reminds us that only by using many pieces of data and subjecting these to proven statistical techniques can you understand how a process is operating and thus make "fact-based" decisions that result in real improvements.

> For more information, see: W. Edwards Deming, *Out of the Crisis* (Cambridge, MA: MIT Center for Advanced Engineering, 1986); Sarv Singh Soin, *Total Quality Control Essentials* (New York: McGraw-Hill, 1992); N. Logothetis, *Managing For Total Quality: From Deming to Taguchi and SPC* (Englewood Cliffs, NJ: Prentice Hall, 1992); PQ Systems, *Total Quality Transformation: Improvement Tools* (Miamisburg, OH: Productivity Quality Systems, 1994); George L. Miller and LaRue L. Krumm, *The Whats, Whys & Hows of Quality Improvement* (Milwaukee, WI: ASQC Quality Press, 1992); Hy Pitt, *SPC for the Rest of Us* (Reading, MA: Addison-Wesley, 1994).

D chart This stands for *demerit chart* (see this entry.)

Death Spiral This biological metaphor is an American slang term meaning a sequence of events that rapidly destroys a company or program. Often the primary symptoms of a company in a death spiral are shrinking product lines, reduced sales force or customer coverage, or layoffs of large numbers of employees. Typically multiple symptoms are present and actually feed on each other as causes of decline.

We can also see these symptoms, however, as wake-up alarms that can help a company become aware of its inefficient processes and poor customer focus. This can lead to the adoption of TQM principles and practices as a way to turn things around. The term is always used, however, to reflect a negative assessment of performance. In analyzing a company in a death spiral, we can use the tools of system management. These help us to understand destructive behavior patterns as a system of reinforcing feedback loops. Understanding patterns using this tool helps a company turn itself around and helps other companies avoid such problems.

> For more information, see: Peter M. Senge, *The Fifth Discipline* (New York: Doubleday Currency, 1990); Peter M. Senge et al. *The Fifth Discipline Fieldbook* (New York: Doubleday Currency, 1994).

Decision Matrix This is a tool that helps in decision making. It lists solutions or problems in cells along the left vertical column and criteria for selection among them in cells along the top horizontal row.

You use this tool by ranking solutions using scales of 1 to 5, 1 to 8, or 1 to 10 depending on how well each solution meets selection criteria. Total scores are added up horizontally to determine which solution is the most attractive. Problem-solving or process teams use such matrices to analyze problems and come up with possible solutions. Figure 30 below shows a decision

J. Willard Marriott commented in 1960 that he hoped the company he inherited from his father could one day be as successful as Howard Johnson. By 1985, Marriott had not only become as successful as Howard Johnson, but had far surpassed it—by a factor of seven times. What happened? While Marriott continued to invest and build for the future, Howard Johnson became overly focused on cost control, efficiency, and short-term financial objectives. While Marriott pushed itself to continually improve the quality and value of its services, Howard Johnson became overpriced and understaffed purveyors of pallid food, hamstrung by outdated ideas.

James C. Collins and Jerry I. Porras, from Built To Last

Solution Alternatives	Effect Least = 1 Most = 5	Cost Most = 1 Least = 5	Time Most = 1 Least = 5	Total
Classroom Training	4	3	2	9
On-the-Job Training	2	4	4	10
Combination Training	5	4	3	12

FIGURE 30

A decision matrix concerning training decisions.

matrix developed by a team for deciding on training questions. The ratings they used indicate that a combination of classroom and on-the-job training might work best.

You should not confuse use of this tool with decision making, which concerns how a group decides to do something (for example, by majority, consensus, or person in charge). However, the use of matrices to organize data is very useful because we can rank alternative solutions and see which among them might best meet selection criteria.

For more information, see: James H. Saylor, *TQM Field Manual* (New York: McGraw-Hill, 1992).

Decision Point In any sampling plan, this is the point at which enough information is accumulated to make a decision concerning acceptance, rejection, or continuing.

Decision points are often highlighted in the documentation of existing or anticipated process steps. In a flowchart, a decision point is indicated by a diamond shape (see *Decision Symbol*). This documentation tells a process owner to figure out what facts are needed to make a decision and what training is necessary for a process user. Decision points can be quite simple. For example, if you are a customer service operator in a utility company and have to decide whether a new customer should make a deposit before you agree to turn on electricity, you would need certain pieces of information (such as previous credit history).

Your decision point would come immediately after checking credit history.

> For more information, see: H. James Harrington, *Business Process Improvement* (New York: McGraw-Hill, 1991); Bruce Brocka and M. Suzanne Brocka, *Quality Management* (Burr Ridge, IL: Irwin Professional Publishing, 1992).

Decision Symbol In a process flowchart, it is a diamond-shaped symbol that indicates that a decision has to be made and often what kind of a decision.

For example, in flowcharting a process describing what we do when we get up in the morning, a decision symbol might be used after the alarm clock rings to indicate that we must decide whether to turn it off and get up or turn it off and go back to sleep for another ten minutes. Figure 31 shows a part of a flowchart with a decision symbol.

> For more information, see: George L. Miller and LaRue L. Krumm, *The Whats, Whys & Hows of Quality Improvement* (Milwaukee, WI: ASQC Press, 1992).

FIGURE 31

A decision symbol, a diamond shape indicating the process can go in one of two directions.

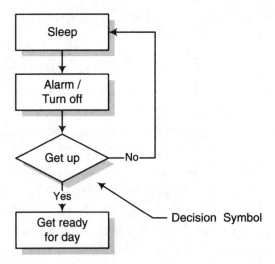

Decreed Quality Standard This is any set of quality standards mandated by circumstances or individuals outside of an organization's control. These standards may be imposed by law, customer requirements on their vendors, society, or management outside of the process team.

Defect This is (1) any nonconformance from a customer's requirement, (2) any attribute of a product or service that fails to meet specifications, (3) any state of unfitness for use.

Defects can be service oriented (for example, the number of unanswered phone calls or missed delivery dates) or faulty products (for example, products that are poorly designed and do not operate or fit properly, or components that do not fit or work). Defects can also come from poorly executed and/or designed processes resulting in finished products or services that, if delivered as designed, would meet customer requirements. Such defects can also come from poor materials, sloppy work habits, or other causes that compromise the final output.

Defects are usually catalogued into four types or classes by degree, from one to four:

1. *Very Serious*: Can cause significant injury or major financial problems.
2. *Serious*: Can cause possible injury or major financial loss.
3. *Major*: Affects the ability of a product or service to do as designed.
4. *Minor*: Includes some problem but still capable of performing.

The elimination of defects remains a core principle of all quality improvement or process work. Statistical process control (SPC) measures of performance of a process are oriented toward understanding, controlling, and eliminating defects. Defects are always seen as deviations from requirements or expectations (targets). At least one writer, Roger Tunks, suggests using the word "defect" instead of "mistake" when talking about results. Mistake focuses attention on who did something wrong rather than more realistically on what about the process caused this unsatisfactory result.

> The word "defect" is a descriptive way of referring to an error, failure, or something broken without placing blame. The word "mistake" draws the question, "Who did it?" Change your language and begin using the term "defect." The change will foster cooperative solutions to problems without placing blame.
>
> *Roger Tunks*

For more information, see: Henry L. Lefevre, *Quality Service Pays: Six Keys to Success!* (Milwaukee, WI: ASQC Quality Press, 1989); Ellis R. Ott and Edward G. Schilling, *Process Quality Control*, Second Edition (New York: McGraw-Hill, 1990); Armand V. Feigenbaum, *Total Quality Control*, Third Edition, Revised (New York: McGraw-Hill, 1991); Eugene L. Grant and Richard S. Leavenworth, *Statistical Quality Control*, Sixth Edition (New York: McGraw-Hill, 1988), John L. Hradesky, *Productivity and Quality Improvement* (New York: McGraw-Hill, 1988); Roger Tunks, *Fast Track to Quality* (New York: McGraw-Hill, 1992).

Defect Elimination   This is the act of identifying and then eliminating defects in products or services arising from process problems. It refers to a group of techniques for reduction of failures or defects in processes.

Some people believe that the major goal of TQM should be the elimination of defects in process outputs, be those products or services. However, W. Edwards Deming and many others have pointed out that defect-free products and services are not sufficient to satisfy customers. Only when firms design and deliver outputs that are both defect-free and deliver customer-desired benefits can the company succeed over the long run. Further, managers must be aware that what customers want is constantly changing. With that said, most firms have a long way to go in the elimination of defects, and that is the main purpose of many of the statistical tools and techniques of TQM.

For more information, see: W. Edwards Deming, *The New Economics* (Cambridge, MA: MIT Center for Advanced Engineering Study, 1993); Ellis R. Ott and Edward G. Schilling, *Process Quality Control*, Second Edition (New York: McGraw-Hill, 1990); Armand V. Feigenbaum, *Total Quality Control*, Third Edition, Revised (New York: McGraw-Hill, 1991); John L. Hradesky, *Productivity and Quality Improvement* (New York: McGraw-Hill, 1988).

Defective   This is a noun that describes a unit that does not conform to requirements.

For example, a computer file becomes a defective when a person keys erroneous data into a computer. For a barber, it could mean cutting a customer's hair too short. In general, it describes a unit of product or service that contains at least one defect. Setting standards for what constitutes a defective or acceptable product is a powerful concept that helps process teams identify existing weaknesses in processes and establish goals or targets for performance. This helps for both incremental or reengineered processes.

For more information, see: Henry L. Lefevre, *Quality Service Pays: Six Keys to Success!* (Milwaukee, WI: ASQC Quality Press, 1989); Roger Tunks, *Fast Track to Quality* (New York: McGraw-Hill, 1992); David A. Garvin, *Managing Quality* (New York: The Free Press, 1988).

Defect Prevention This refers to the approaches involved in eliminating process problems that result in defective outputs.

These approaches all focus on eliminating such problems as close to their source as possible rather than depending on inspection after the fact. For example, consider a company that produces computer keyboards. At the end of the line, the inspector discovers one key is sticking. There are three alternatives: (1) Fix the defectives, (2) take them apart and reuse the materials, or (3) discard them. All of these are expensive and waste time and materials. With an emphasis on defect prevention, the company would quickly use a variety of tools to pinpoint the source of this problem, such as an improperly adjusted die, and make adjustments. However, rather than waiting for such problems to appear, defect prevention urges continuous checking at each step and catching process errors as soon as they become evident at any step in the process. (See also *Prevention.*)

For more information, see: William Lareau, *American Samurai* (New York: Warner Books, 1991).

> Mistakes are made, and defects are produced, at the micro-process level. Every manufacturing defect (and every paper/data process defect) can be traced to a specific action or set of actions. If these actions are changed, the defect will not occur.
>
> *William Lareau*

Defects Per Million This is the number of defects (errors) that occur in one million opportunities.

This type of measure is widely used in large companies and is applicable across all actions of a process, product, or service; hence its popularity. A popular variation is the goal of Six

Sigma, which has come to signify 3.4 errors per million items. Thus, Six Sigma, as used by Motorola (the company with which it is most identified), has come to mean that about 999,997 times out of a million its outputs will be defect free. Motorola's adoption of Six Sigma encouraged many other companies to do the same. The one weakness with the measure is that nontechnical employees sometimes have difficulty with the concept. (See also *Six Sigma.*)

For more information, see: Joseph M. Juran and Frank M. Gryna, *Quality Planning and Analysis*, Third Edition (New York: McGraw-Hill, 1993).

Deficiency This is a failure in a product or service that causes customer dissatisfaction. The same idea also applies to process performance. Users or customers are the ones who identify deficiencies.

Demerit Chart This is a tool for tracking defects of various types in process outputs. This type of chart is one more way to measure the quality of outputs from a process.

Demerits are usually broken down into the four categories of defects: critical, serious, major, and minor. These are given weights, for example 100 for critical, 25 for serious, 10 for major, and 1 for minor. Defects of each type are then tracked for a specified number of units, such as a lot. If the number of units were 500, the demerit chart might look as shown in figure 32. Using this total, we can now divide the total number of demerits by the

FIGURE 32
A typical demerit chart.

Type of Defect	Number Found	Demerits per Defect	Total Demerits
Critical	2	100	200
Serious	5	25	125
Major	16	10	160
Minor	47	1	47
Total	70		532

total number of products (532/500), which equals 1.06 demerits per unit. Using this as a baseline, the company can then track additional lots or batches of outputs to measure improvement.

> For more information, see: Armand V. Feigenbaum, *Total Quality Control*, Third Edition, Revised (New York: McGraw-Hill, 1991); Joseph M. Juran and Frank M. Gryna, *Quality Planning and Analysis*, Third Edition (New York: McGraw-Hill, 1993).

Deming Cycle This is another term for the plan-do-check-act (PDCA) improvement cycle. Deming always called it the "Shewhart cycle," after Walter Shewhart, who originated it. It was Deming, though, who made it popular.

The term is not as widely used as is PDCA. It is a basic strategy for continuously improving any process and the outputs of that process. Most practitioners of process improvement use some form of PDCA. (See *Continuous Improvement* for further discussion.)

> For more information, see: Peter R. Scholtes, *The Team Handbook* (Madison, WI: Joiner Associates, 1988); W. Edwards Deming, *Out of the Crisis* (Cambridge, MA: MIT Center for Advanced Engineering Study, 1986); Brian L. Joiner, *Fourth Generation Management* (New York: McGraw-Hill, 1994); and many other books that discuss basic quality improvement techniques.

> The perception of the cycle [for improvement] came from Walter Shewhart. I called it in Japan in 1950 and onward the Shewart cycle. It went into immediate use in Japan under the name of the Deming cycle, and so it has been called there ever since.
>
> *W. Edwards Deming*

Deming Prize Established in 1951 by the Union of Japanese Scientists and Engineers (JUSE), it recognizes companies that have made outstanding progress in improving the quality of their operations.

This very important quality prize is named after W. Edwards Deming who, in 1950, first introduced the Japanese to many industrial quality control practices. After Deming gave his lectures in 1950, they were bound into a book and sold. Rather than accept royalties, Deming suggested that JUSE keep them to encourage the implementation of these techniques, and the royalties were used to fund the prize.

This is Japan's major quality award. It combines many of the characteristics of Deming's improvement philosophy and strategy. It is awarded annually and recognizes the superior application of *company-wide quality control* (see this entry) based on statistical quality control. Companies that apply for this award are judged in ten key categories:

1. Policy and objectives.
2. Organization and its operation.
3. Education and its extension.
4. Assembling and disseminating information and its utilization.
5. Analysis.
6. Standardization.
7. Control.
8. Quality assurance.
9. Effects (results).
10. Future plans.

Besides being awarded to companies or divisions or factories of companies, the Deming Prize is also awarded to individuals who have made outstanding contributions to the study and/or dissemination of statistical methods for CWQC. Though most of the companies that apply for the prize are Japanese, any company can apply. A prominent American winner in 1987 was Florida Power & Light.

For more information on the Deming Prize, write to: The Union of Japanese Scientists and Engineers (JUSE), 5-10-11 Sendagaya, Shibuya-ku Tokyo 151, JAPAN.

Dependability This is the degree to which a product or service functions and remains capable of performing as required by specifications. It is related to the concept of reliability.

Dependability is the probability that a device will work as it is supposed to. If car batteries of a certain brand are said to be 90 percent dependable over five years, this means that 90 percent of these batteries will last at least five years.

For more information, see: Dev G. Raheja, *Assurance Technologies* (New York: McGraw-Hill, 1991); Armand V. Feigenbaum, *Total*

Quality Control, Third Edition, Revised (New York: McGraw-Hill, 1991).

Deployment This concept refers to the actions taken to implement a program or any consistent management approach across the entire organization or over multiple departments.

The idea of deployment of quality management principles and practices is a central issue in judging for the Baldrige Award. In the Baldrige Criteria, deployment addresses the extent to which a company has implemented and uses process and quality practices as standard operating procedure.

Another idea implied by deployment is the particular set of actions a company takes to implement a set of goals. For example, deployment of TQM could refer to how management introduces employees to quality principles, process reengineering, and statistic process control. It could also refer to the tactics management employs to change the culture of an organization.

For more information, see: Thomas H. Berry, *Managing the Total Quality Transformation* (New York: McGraw-Hill, 1991); David A. Garvin, *Managing Quality* (New York: The Free Press, 1988); Stephen George, *The Baldrige Quality System* (New York: John Wiley & Sons, 1992).

Deployment Matrix This is a structured method for translating customer requirements into specific characteristics of a product or service. The idea is to take general requirements, such as non-breakability for the case of a radio, and turn that into a specific measure of plastic strength. It might also be used in services to go from the general idea of customer satisfaction to specifics of what characteristics of the service will generate satisfaction. This type of matrix is similar to the matrices used in *quality function deployment* (see that entry).

For more information, see: Eugene H. Melan, *Process Management: Methods for Improving Products and Services* (New York: McGraw-Hill, 1993); Bruce Brocka and M. Suzanne Brocka, *Quality Management* (Burr Ridge, IL: Irwin Professional Publishing, 1992).

Designing In Quality vs. Inspecting In Quality This captures the notion of prevention versus detection. Designing in quality emphasizes that quality should be a concern right from the beginning of the design process to avoid problems down the road. It raises management consciousness that it is more efficient and effective to improve quality at the design phase to minimize costs involved with inspections down the line. In other words, the goal is to eliminate the causes of poor quality by anticipating problems early. This includes creating designs that are simple and robust. This means the item will be easier to assemble and that the designs will minimize failure of parts. Inspecting in quality is a more traditional and costly approach to guaranteeing quality. It is an approach that emphasizes the necessity of rework and waste to get rid of defectives, not the prevention of defectives in the first place.

Here is an example of designing in quality: We can look at how a utility company might design a process for hooking up new customers' electricity. In doing this, a process owner and team members will want to ensure that the customer's name, address, and telephone number are correct. An error here might cause a utility truck to go to the wrong house to do the electrical hook-up. Therefore, to avoid that problem downstream in the process, the team would design in a method for how the customer service representative collects and records new customer information. They would then develop a process for eliminating errors at this stage. They might incorporate in this process a step that calls for the operator to read back to a prospective customer the newly acquired information on name, address, and telephone to verify its accuracy. While this example may seem simple, many problems in processes happen just because team members have not undertaken such an analysis and examination of problem sources. (See also *Robust Design* and *Poka-Yoke*.)

For more information, see: Eugene R. Carrubba and Ronald D. Gordon, *Product Assurance Principles* (New York: McGraw-Hill, 1988); William Lareau, *American Samurai* (New York: Warner Books, 1991); Ernst & Young, *Total Quality: An Executive's Guide*

SPC is a preventive measure, but only to a point. It addresses only the variation owing to the many variables in the production process itself. In a broader sense, it is still reactive and represents after-the-fact problem solving. Prevention and reduction of variation, however, can and should begin much earlier in the life of product—with design.

Ronald M. Fortuna from Total Quality: An Executive's Guide for the 1990s *for the 1990s* (Burr Ridge, IL: Irwin Professional Publishing, 1990).

Design of Experiments (DOE) This is a formal branch of applied statistics that focuses on planning, conducting, studying, and interpreting the results of controlled tests. This is a common approach for figuring out how to improve a process. Genichi Taguchi, a Japanese quality expert, is a pioneer in developing the DOE approach.

DOE provides a methodology for scientifically studying the factors that affect variation in process outputs. This helps us come up with the right combination of factors that will yield results consistent with specifications and reduce variation. Using DOE we can find out which factors affect results in a positive or negative manner and which ones have little affect on results. We can contrast DOE with the traditional approach to dealing with a process problem, which is to "check this, do this, check that," in a trial and error fashion. This approach is just as likely to make things worse as it is to improve them. And it provides no real explanation of what combination of factors works best or why.

Common steps in DOE are to (1) define a problem to be solved (often the output of a process does not meet specifications); (2) list the factors that might affect the way the process operates (for example, temperature, steel hardness, tool speed, lubrication, if we are talking about a machining process); (3) conduct experiments that allow for different combinations of these factors to be studied; (4) choose the combination that yields the best result. You can then document the process according to your results and train employees in the new method.

The DOE technique is widely used in manufacturing. It is becoming increasingly popular in all types of business activities in which there are alternative methods or approaches, and managers want to know which of these would work best. An example might be in direct mail, where the experimenter tries various

> A typical designed experiment can be described as a box with several knobs, called *factors*, on one side and a meter on the other side, which displays the results of turning the knobs. A typical objective of designed experiments might be to determine which factor (knob) produces the greatest changes in output (meter).
>
> *John L. Hradesky*

combinations of appeals and mailing lists to determine which combination pulls the most responses.

For more information, see: Phillip J. Ross, *Taguchi Techniques for Quality Engineering* (New York: McGraw-Hill, 1988); John L. Hradesky, *Productivity and Quality Improvement* (New York: McGraw-Hill, 1988); William Lareau, *American Samurai* (New York: Warner Books, 1991).

Design Phases These are three specific phases in the design of a process or product: systems design, parameter design, and tolerance design. These phases were designated by Genichi Taguchi, a leading Japanese quality expert.

Engineers frequently employ this phase approach in the design and manufacture of products, such as in the development of new computer chips. It depends heavily on the use of statistical process control techniques for gathering and analyzing data at each phase.

For more information, see: Phillip J. Ross, *Taguchi Techniques for Quality Engineering* (New York: McGraw-Hill, 1988).

Design Review This a formal process for detecting problems during the design of new products or new services. It usually involves a design review team of people whose work will be impacted by the design. This team can include individuals from manufacturing, production, marketing, maintenance, and even customers. They meet to review design ideas and provide feedback so that the final design minimizes problems in each of these areas and meets customer requirements. We should note that customer participation in this process is encouraged by the Baldrige Criteria.

For more information, see: Joseph M. Juran and Frank M. Gryna, *Quality Planning and Analysis,* Third Edition (New York: McGraw-Hill, 1993); Joseph M. Juran, *Juran on Planning for Quality* (New York: The Free Press, 1988); Eugene R. Carrubba and Ronald D. Gordon, *Product Assurance Principles* (New York: McGraw-Hill, 1988).

Detailed Process Diagram This is another term describing a flowchart documenting each step in a process.

This is the most widely used tool for documenting either an existing process or a proposed new one because it is one of the easiest techniques available for describing how a process works. It can be as simple as illustrated in figure 33, in which only the basic steps are defined, or more detailed as in figure 34, sometimes called a deployment flowchart, which shows who participates in which steps.

FIGURE 33

Simple flowchart of book production process.

FIGURE 34

A deployment flowchart for part of the book production process.

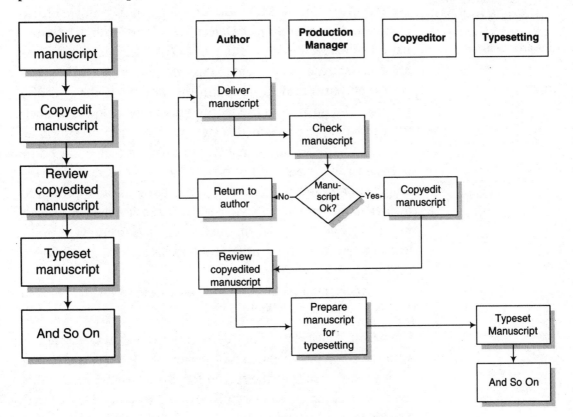

The advantages of using process diagrams are (1) everyone can see the steps involved, (2) it becomes easier to discuss gaps in time and performance from step to step, (3) it is a clear shorthand overview that can be supported by pages of narrative describing the steps in more detail. The approach was borrowed from computer programmers, who for decades have used flowcharting methods to define all the steps in a program.

For more information, see: H. James Harrington, *Business Process Improvement* (New York: McGraw-Hill, 1991); George L. Miller and LaRue L. Krumm, *The Whats, Whys & Hows of Quality Improvement* (Milwaukee, WI: ASQC Quality Press, 1992); and many other books on quality management tools.

Detection This is the act of identifying any nonconformance to some standard *after the fact.* Detection is done through a combination of inspection and testing to find defects before products are declared ready to present to customers.

Employees perform detection on outputs at the end of a process. The purpose is to identify unacceptable levels of quality. Most people think of this step as inspection of a product coming off the assembly line (for example, a soft drink bottle only half full) or an audit of how something was done (for example, an accounting procedure). Detection as an approach to quality management is losing popularity in favor of in-process inspections and designed-in quality that seek to prevent problems altogether or to identify problems as early as possible in a process. (See also *Prevention.*)

For more information, see: William Lareau, *American Samurai* (New York: Warner Books, 1991).

Deviation This refers to any departure from a set of desired or anticipated values or pattern of performance. Standard deviation is the measure of the spread for most distributions and is used primarily in monitoring process performance. Deviation is also defined as the expected spread in the performance of a process from some standard or average. (See also *Standard Deviation.*)

Deviation is used whenever someone wants to measure variability in a process. It helps you understand what percent of outputs will vary by what percent from the target specification or

Constructing flowcharts disciplines our thinking. Comparing a flowchart to the actual process activities will highlight the areas in which rules or policies are unclear or are even being violated. Differences between the way an activity is supposed to be conducted and the way it is actually conducted will emerge.

H. James Harrington

process mean. In process improvement work, an important goal is to reduce the size of the standard deviation from specification.

> For more information, see: Ellis R. Ott and Edward G. Schilling, *Process Quality Control*, Second Edition (New York: McGraw-Hill, 1990); George L. Miller and LaRue L. Krumm, *The Whats, Whys & Hows of Quality Improvement* (Milwaukee, WI: ASQC Press, 1992).

Diagnostic Journey and Remedial Journey Used together, this is a two-part investigatory process employed by teams to solve chronic quality problems. In the diagnostic phase a team will go from identification of a symptom to its cause. In the second, it moves from cause to remedy.

Breakthrough to quality improvement, to use Joseph M. Juran's approach, involves several basic steps that range from identifying the need for change to identification of a project, followed by organization for making improvements, and then a series of steps designed to diagnose problems and areas for improvement. These are followed by steps taken to remedy problems. Juran argues that these steps lead to a breakthrough in resistance to change within an organization followed by holding on to the achieved gains. The key to the diagnostic phase is to appreciate that 80 percent or more of all defects are the result of how processes operate. These can be addressed only by improving the system's processes. This further suggests that only 15 to 20 percent of problems are operator controlled. Thus, diagnosis and remedy should mainly focus on process analysis and improvement using the tools of TQM, such as data collection devices, SPC charts, and cause-and-effect diagrams.

> For more information, see: Joseph M. Juran, *Juran on Leadership for Quality* (New York: The Free Press, 1989); Joseph M. Juran and Frank M. Gryna, *Quality Planning and Analysis*, Third Edition (New York: McGraw-Hill, 1993).

Diagnostic Procedure These are the actions taken to discover the causes of quality deficiencies in products, processes, and services.

For more information, see: Joseph M. Juran and Frank M. Gryna, *Quality Planning and Analysis,* Third Edition (New York: McGraw-Hill, 1993).

Directed Survey This is the opposite of a blind survey. Participants know who the sponsor is and why the sponsor is asking questions. It is a standard approach when asking customers for suggestions on how to improve products or services.

Commonly used by vendors of many different kinds of products, it leads to specific information about how customers use these products, what problems they encounter, and how the products and accompanying services can be improved. This technique can be employed in a number of ways: mail surveys, telephone surveys, and focus groups. Mail surveys allow you to get to many people quickly, but the return rates are quite low. Telephone surveys yield much higher returns, but are expensive. Focus groups yield the greatest amount of information, but limit the number of customers you contact. Information is summarized both in numeric and textual forms.

For more information, see: Bob E. Hayes, *Measuring Customer Satisfaction: Development and Use of Questionnaires* (Milwaukee, WI: ASQC Quality Press, 1992); Richard C. Whiteley, *The Customer-Driven Company* (Reading, MA: Addison-Wesley, 1991).

Disciplined Continuous Improvement Methodology This is another, formal, way of indicating commitment to continuous quality improvement. It is associated with the Department of Defense.

For more information, see: James H. Saylor, *TQM Field Manual* (New York: McGraw-Hill, 1992.

Distributed ABC This represents a company-wide commitment to activity-based costing (ABC) and making available ABC information to everyone in the company to assist in their improvement efforts. It suggests that ABC has the most value when everyone has access to and can use this information to better understand and improve their efforts in relation to others in the delivery of outputs to customers.

For more information, see: Peter B.B. Turney, *Common Cents: The ABC Performance Breakthrough* (Portland, OR: Cost Technology, 1991).

Distribution In statistical terms, this describes the amount of potential variation in outputs of a process that is under statistical control. In a common cause system, you cannot predict the actual amount of variation in individual outputs, though you can predict the range of variation. That range is the distribution of points. It is the area found under a typical bell-shaped curve or within the control limits on a statistical process control chart. Another graphical presentation of distributions is shown in a *histogram*. (See this term.)

For more information, see: George L. Miller and LaRue L. Krumm, *The Whats, Whys & Hows of Quality Improvement* (Milwaukee, WI: ASQC Quality Press, 1992); Joseph M. Juran and Frank M. Gryna, *Quality Planning and Analysis*, Third Edition (New York: McGraw-Hill, 1993).

Documentation This is all the documents that explain how a process operates and what needs to be done to properly manage, execute, and improve the process.

Such documentation is valuable as a baseline for understanding current processes and thus making improvements. It is also very valuable for training new employees and maintaining consistency in the way processes are executed. A big problem in many companies is not having documentation for their processes. This leads to confusion, lack of coordination, long lead times for new employees to learn, and inconsistency among employees in getting things done.

For more information, see: H. James Harrington, *Business Process Improvement* (New York: McGraw-Hill, 1991).

Documentation Change Control This is any management system for preventing unauthorized changes to documents or process documentation and for making changes in a controlled manner.

Well-run projects, such as process improvement and software programming, pay close attention to documentation.

This suggests the need for a process to properly maintain this documentation and make changes to it in a specified and agreed upon manner. Project management techniques usually include a numbering system to track revisions to documents, policies for signing them out by project members, and standardization of format for content, cataloguing, and modification. Lack of attention to documentation controls can be a major source of project and process problems. It also makes it difficult to assess problems and make improvements.

For more information, see: H.G. Menon, *TQM in New Product Manufacturing* (New York: McGraw-Hill, 1992);

Document Symbol In a flowchart, it is any symbol that denotes a document that is part of the process, such as a purchase order, a proposal, or some other necessary document that moves the process forward. Figure 35 illustrates the symbol for a document.

FIGURE 35

Part of a process flowchart showing a document symbol.

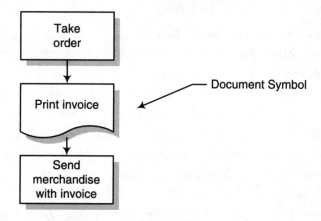

For more information, see: George L. Miller and LaRue L. Krumm, *The Whats, Whys & Hows of Quality Improvement* (Milwaukee, WI: ASQC Quality Press, 1992); and several other books that discuss quality improvement tools.

Dodge-Romig Sampling Plan Tables Developed by Harold F. Dodge and Harry G. Romig, these are four sets of tables: single-sampling lot tolerance, double-sampling lot tolerance, single-sampling average outgoing quality limits, and double-sampling average outgoing quality limit tables. Based on the sampling plan used, these tables allow vendors and customers to determine the probability that any lot or batch of products will meet requirements for quality.

The tables indicate, for different sampling plans and for different acceptable quality levels (that is, what percent of defects would make a lot acceptable or not), the possibility that 10 percent of the time, the plan would let more defectives through than would be acceptable. In other words, these tables help companies decide which plan will give them 90 percent reliability that any lot or batch of products will be of acceptable quality. (See also *Operating Characteristic Curve.*)

> For more information, see: Henry L. Lefevre, *Quality Service Pays: Six Keys to Success!* (Milwaukee, WI: ASQC Quality Press, 1989); Ellis R. Ott and Edward G. Schilling, *Process Quality Control,* Second Edition, Second Edition (New York: McGraw-Hill, 1990); Eugene L. Grant and Richard S. Leavenworth, *Statistical Quality Control,* Sixth Edition (New York: McGraw-Hill, 1988).

DOE This is the abbreviation for *design of experiments* (see this entry).

Dominance This is an often-employed strategy by customers to insist upon their will toward a supplier. Customer dominance happens when there are numerous options for sourcing supply of goods, materials, and services. Conversely, when a monopoly situation exists, then the supplier becomes the dominant party in the relationship.

In either situation, dominance represents "old thinking" in how suppliers and customers should relate to each other. A more quality-focused approach is to treat each other as equals and respect each other's needs to form a partnership that improves the overall performance of both.

A second definition of dominance involves the role of a variable in a process. In this instance, among a number of variables there may be one that dominates. In other words, it exerts greater influence over the performance of the process than any other or even all the others combined. In manufacturing, common instances of dominance include setup-dominance, component-dominance, and time-dominance. In services, there are others that also occur in manufacturing: information-dominance and worker-dominance.

> For more information, see: W. Edwards Deming, *Out of the Crisis* (Cambridge, MA: MIT Center for Advanced Engineering Study, 1986); William Lareau, *American Samurai* (New York: Warner Books, 1991).

Double Sampling This is an inspection technique in which you inspect a first lot of n1 size, which leads you either to accept or reject it. If you reject it, then you inspect a second sample of a larger lot size n2, which in turn leads to a decision to accept or reject the lot. It is a way of checking your sampling technique to make sure you are selecting a lot of sufficient size to detect whether a process is delivering the percentage of acceptable outputs consistent with a specification.

> For more information, see: Joseph M. Juran and Frank M. Gryna, *Quality Planning and Analysis*, Third Edition (New York: McGraw-Hill, 1993)

Downstream Value Chain These are the linked processes by which value is added step-by-step to deliver quality to customers and sustain organizational profitability.

For example, a distributor or dealer who sells your company's products may be considered part of the downstream value chain for your products. The concept is an extension of the value chain model first introduced by Michael Porter. He said all activities of a company that are intended to "design, market, deliver, and support its product [or service]" can be depicted in his now famous value chain chart. Each activity has a value that can be understood. Business people and Porter have extended this model to include value-added activities that occur outside of the manufacturing enterprise that are required to complete the flow of activities resulting in the sale

> A firm's value chain and the way it performs individual activities are a reflection of its history, its strategy, its approach to implementing its strategy, and the underlying economics of the activities themselves.
>
> *Michael Porter*

PRIMARY ACTIVITIES

SUPPORT ACTIVITIES

and delivery of a product or service to a customer. Properly managing the value chain results in a company making its profit margin.

For more information, see: Michael E. Porter, *Competitive Advantage: Creating and Sustaining Superior Performance* (New York: The Free Press, 1985); Brian L. Joiner, *Fourth Generation Management* (New York: McGraw-Hill, 1994).

Downtime A slang term originating in information processing, it is the time during which a machine or process is not functioning due to a failure of equipment or some step in the process. A planned shutdown to perform maintenance is also a cause of downtime.

Maintenance is now considered a major process for any organization that uses computers, trucks, and other types of equipment upon which many employees depend. Companies can track unplanned downtimes, discover the root causes for this happening, and deploy methods to bring about higher levels of availability (the opposite of downtime). Downtime is usually considered a failure or weakness in a process. Minimizing downtime is one important aspect of reducing cycle time and operating costs.

For more information, see: G. Gordon Schulmeyer and James I. McManus (eds.), *Total Quality Management for Software* (New York: Van Nostrand Reinhold, 1992); George Stalk, Jr. and

FIGURE 36

A generic value chain. Managing each of these processes is key to satisfying customers and to profitability. Adapted from Michael Porter, *Competitive Advantage* (New York; The Free Press, 1985).

Thomas M. Hout, *Competing Against Time* (New York: The Free Press, 1990); Masaji Tajiri and Fumio Gotoh, *TPM Implementation* (New York: McGraw-Hill, 1992).

Drift This refers to the measurements of a process that do not exhibit a change in variabililily but in which there is a gradual shift in the average performance. Drift can result in a shift of measurable results of a machine due to wear and tear on components. It can be caused by a gradual change in the performance of a process due to some special cause being gradually repeated many times.

We can identify drift by careful measurement of process performance and monitoring whether performance is within control limits. If a pattern in a particular direction is developing, this indicates drift is occurring. Drift is considered a malperformance in both equipment and processes. Unchecked, fundamental performance will deteriorate, that is to say, fail to conform to requirements. A mean chart will usually help a process owner to recognize drift. As part of the analysis of special causes, what usually is done is to incline control limits in the direction of the drift. Then the extent of the drift can be calculated. Figure 37 below shows a process drift with upward sloping control limits.

> Quality troubles often gradually "drift" into a process. An improperly ground tool may cause a trend toward unusual variation which will finally result in the production of nonconforming parts.
>
> *Armand V. Feigenbaum*

FIGURE 37

An example of process drift, with control limits shown going upward.

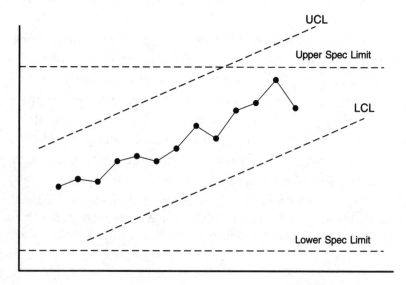

For more information, see: N. Logothetis, *Managing For Total Quality: From Deming to Taguchi and SPC* (Englewood Cliffs, NJ: Prentice Hall, 1992); Armand V. Feigenbaum, *Total Quality Control*, Third Edition, Revised (New York: McGraw-Hill, 1991).

Dry Run This is any test of a process, software, or procedure under actual operating conditions before putting it into actual use.

Early Warning More than just an advance sign that something is wrong with the performance or design of a process, product, or service, it is a mechanism for getting customer feedback before a project or product is completed.

The best practice is to involve customers in reviewing the design of products and services early on to understand if your approach will create problems for them or earn their disapproval prior to doing any significant development work.

> For more information, see: Roy A. Bauer, Emilio Collar, and Victor Tang, *The Silverlake Project* (New York: Oxford University Press, 1992).

Economic Quality This is the level of quality at which costs exceed the extra benefits the product provides. It is another way of discussing the notion of diminishing returns. It also reminds managers that customers define what is quality to them by weighing costs versus benefits.

Effect This concept is often applied to a problem or defect that is the result of some flaw or inefficiency within a process or activity. This term is often used in conjunction with the results of design of experiments. Changing different factors in an experiment yields different effects.

For example, an experiment might be devised to measure the effect of doing something to a product on one machine versus another (for example, cut metal). Statistical analysis of experiments of new steps in a process can lead to an assessment of the effects of a step to a process (for example, measuring the average effects of an operator's performance across various machines). The idea of effect is also central in the use of a cause-and-effect diagram quality tool.

> For more information, see: Ellis R. Ott and Edward G. Schilling, *Process Quality Control*, Second Edition (New York: McGraw-Hill, 1990).

Effective The degree to which the output of an organization meets its own goals or, more particularly, the standards of performance required by customers. Peter Drucker called it "doing the right things."

No matter how well you manage a process, unless the outputs from that process deliver benefits that customers will value, you are not effective.

> For more information, see: Peter F. Drucker, *Management: Tasks, Responsibilities, Practices* (New York: Harper & Row, 1973).

Effective Process This phrase refers to a process that generates results that conform to customer requirements or to those of the process owner. Lack of effectiveness is the difference between actual and required performance, usually expressed in numeric terms. The difference is always the degree to which results (output) diverge from requirements.

Effectiveness is more than impressions; properly measured, we can identify effectiveness or lack of it using statistics. Common characteristics of an effective process include:

- Having a systematic measurement scheme in place that clearly indicates that customer requirements are being met consistently and are continuously improving (simpler, more efficient, faster, more responsive).
- Sound documentation.
- A clear understanding of its cost and time of operation.

> Even the most efficient business cannot survive, let alone succeed, if it is efficient in doing the wrong things, that is, if it lacks effectiveness. No amount of efficiency would have enabled the manufacturer of buggy whips to survive.
>
> *Peter F. Drucker*

- Routine benchmarking of the process, with this influencing its continuous modification.

- Addressing process problems in a disciplined manner.

For more information, see: H. James Harrington, *Business Process Improvement* (New York: McGraw-Hill, 1991).

Efficiency It is the degree to which costs have been minimized in the production of some output. It is the ratio of the quantity of resources consumed to output delivered. To become more efficient, firms either can deliver more output for the same amount of input or the same amount of output with less inputs. It is also often seen as the ratio of the quantity of resources expended to meet customer requirements versus resources planned.

Efficiency is a way of talking about the productivity of an enterprise. Peter Drucker says it is "doing things right." Ultimately, the heart of TQM is about providing direction for organizations to become ever more efficient and effective.

For more information, see: Peter F. Drucker, *Management: Tasks, Responsibilities, Practices* (New York: Harper & Row, 1973); H. James Harrington, *Business Process Improvement* (New York: McGraw-Hill, 1991).

80/20 Rule Also known as the Pareto principle, it was rescued from obscurity by Joseph M. Juran in 1950 to suggest that most effects come from few causes; in other words, 80 percent of all effects come from only 20 percent of possible causes.

One of the basic tools used in determining the causes of problems is the Pareto chart, a bar graph that lists problems by type and by the number in which they occur. Figure 38 shows a typical Pareto chart.

The assumption is that in a given period, you will find most problems are of one or two types, and that fixing those that occur most frequently will reduce dramatically the total number of problems (hence the 80/20 rule that 80 percent of problems are the result of 20 percent of the causes). By addressing those 20 percent of causes, you can create leverage in eliminating

> Effectiveness is the foundation of success—efficiency is the minimum condition for survival *after* success has been achieved.
>
> *Peter F. Drucker*

> The Pareto principle is a universal for sorting any conglomerate mixture into two neat piles: the vital few and the trivial many.
>
> *Joseph M. Juran*

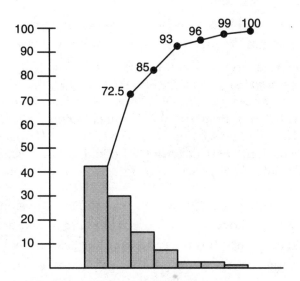

FIGURE 38
A typical Pareto chart
showing that most
problems are similar
in nature.

many problems. Juran calls the difference between the 20 percent of events that cause 80 percent of the effects and the 80 percent of events that cause 20 percent of the effects, "the vital few
and the trivial many."

For more information, see: Michael Brassard and Diane Ritter,
The Memory Jogger II (Methuen, MA: GOAL/QPC, 1994); Joseph
M. Juran, *Managerial Breakthrough*, Revised Edition (New York:
McGraw-Hill, 1994).

Employee Involvement (EI) This is a code term for those management practices in which employees routinely play a critical
role in making decisions about operations, setting priorities,
making and implementing suggestions for improvements, planning, setting goals and targets, and monitoring performance. It
is a major component of modern quality practices. The idea of
EI is often related to the idea of employee empowerment.

It has become a major component of recent corporate
strategies to downsize by eliminating layers of management. To
do that, and still perform functions of the enterprise, senior
management has had to get employees more involved in making
day-to-day decisions that at one time were made by middle
management. Effective employee involvement or empowerment

Employee Involvement
works, but only as a way of
running a business. It is not
something you dabble in. It
does not work as a short-
term solution or an intriguing experiment. It is truly a
Pandora's box, tempting to
open but powerful enough
to change your "world"
forever.

*Stephen George
and Arnold Weimerskirch*

calls for redefined jobs, possibly better pay, and more training in those areas once the private reserve of middle managers.

For more information, see: Kiyoshi Suzaki, *The New Shop Floor Management* (New York: The Free Press, 1993); Jack D. Orsburn et al., *Self-Directed Work Teams: The New American Challenge* (Burr Ridge, IL: Irwin Professional Publishing, 1990); Charles C. Manz and Henry P. Sims, Jr., *Business Without Bosses* (New York: John Wiley & Sons, 1993); Stephen George and Arnold Weimerskirch, *Total Quality Management* (New York: John Wiley & Sons, 1994).

Empowerment This is the act of delegating responsibility and authority to employees to make decisions and to take action. In the world of quality, this is a code word for delegation.

It embraces the assumption that employees, when properly trained, can often make the best decisions on how to manage and improve a process. For empowerment to take hold, it requires a substantial and sustained commitment to a culture that values the contribution of all employees and a management that will not and cannot rescind such authority once given to employees.

Empowerment represents a major change in how many organizations are performing. There is growing evidence that when properly done, better decisions are being made quicker. Keys to success include a clear vision of where the enterprise is going, how it is doing, a good definition of goals and measure, proper use of team-based clusters of employees, and a healthy investment in training. Empowerment also suggests a culture characterized by open communication and the implementation of information technology, which makes information and the ability to consult with others easy and commonplace.

Brocka and Brocka (see references) list the following practices of managers who seek to make empowerment an important part of the way their companies:

- *Foster ownership.* This suggests that management has recognized that employees who have responsibility for processes, projects, and tasks must also have the authority to exercise this responsibility.

[One person's discovery of what empowerment means] She began with the first three steps to Zapp. *1. Maintain self-esteem. 2. Listen and respond with empathy. 3. Ask for help in solving problems.* She started practicing them in all her dealings with people. Then she applied the soul of Zapp: *Offer help without taking responsibility.*

William Byham and Jeff Cox, from ZAPP: The Lightning of Empowerment

- *Value all contributions.* It is important to appreciate that everyone in the organization has something to contribute, and to create an environment where each feels free to make his or her contribution.
- *Listen to the least voice.* You never know from whom the answer to a problem might come. Again, create an environment where this is a core value.
- *Allow teams to own problems.* Management must give teams autonomy to investigate and solve the problems they have taken on. If this does not happen, then there is no reason to create such teams.
- *Delegate authority to the lowest possible organizational level.* Managers should always work at giving the people who have been hired to do jobs the authority and responsibility to do those jobs (and encourage learning from mistakes).

For more information, see: Bruce Brocka and M. Suzanne Brocka, *Quality Management* (Burr Ridge, IL: Irwin Professional Publishing, 1992); Jack D. Orsburn et al., *Self-Directed Work Teams: The New American Challenge* (Burr Ridge, IL: Irwin Professional Publishing, 1990); William C. Byham and Jeff Cox, *ZAPP: The Lightning of Empowerment* (New York: Fawcett Columbine, 1988); Charles C. Manz and Henry P. Sims, Jr., *Business Without Bosses* (New York: John Wiley & Sons, 1993); Stephen George and Arnold Weimerskirch, *Total Quality Management* (New York: John Wiley & Sons, 1994).

Enabler This is any process, technique, or other activity that makes possible the implementation of a process, technique, best practice, or other action. It is always one of the critical success factors making some result possible.

An enabler might be a manager who provides resources to get something done, a facilitator/coach who helps a process team do its work well, a tool (such as software) that makes the completion of the task possible. In process reengineering, information technology often becomes the enabler of dramatic improvement. Other common enablers include changes in organizational culture and the management processes that encourage employee involvement and empowerment.

> For more information, see: Thomas H. Davenport, *Process Innovation: Reengineering Work Through Information Technology* (Cambridge, MA: Harvard Business School Press, 1993); Stephen George and Arnold Weimerskirch, *Total Quality Management* (New York: John Wiley & Sons, 1994).

Engineering Change (EC) This is a documented modification to specifications or to operational methods, materials, or procedures that a company uses to produce an output. Usually such changes are made to enhance the efficiency and/or effectiveness of the process or a product.

> For more information, see: J.R. Taylor, *Quality Control Systems* (New York: McGraw-Hill, 1989).

Engineering Study This is any study and analysis of information using scientific methods of research to define or enhance the performance of a process.

Enterprise Processes Also called business processes, these are the major high level processes of an organization. For example, in a manufacturing company, they could be design, product manufacture, sales, and delivery. These are typically "macro-processes" that catalog the major activities that support the mission of an organization and that become the basis of significant process identification and improvement. Each has many subprocesses, but when added together constitute the major activities of an organization that affect quality.

Organizations are finding that it makes sense to think of their tasks as groups of processes. The task of management is then to coordinate these processes within and across functions. Some examples of enterprise processes include: Development, Distribution, Financial Accounting, Financial Planning, Information Systems, Production Control, Purchasing, Human Resources, and Programming. Each of these has many subprocesses associated with it.

For more information, see: H. James Harrington, *Business Process Improvement* (New York: McGraw-Hill, 1991); James W. Cortada, *TQM For Sales and Marketing Management* (New York: McGraw-Hill, 1993).

Event In statistics or quality control, this is a single step executed one time in a process. A measurement of the output of this event yields a single data point on a quality control chart. These measurements might be the time, closeness to specification, cost, sales dollars.

TQM points out the flaw of managing events rather than processes. Managers who fail to understand statistical process control often take an event as an indication that something good or bad has happened, as for example when sales dip for a month or costs rise, or there are more defects than expected in one sample from several lots of components. Misunderstanding that such events are usually brought about by variation in a process that is under control, managers have a tendency to treat single events as if they were a trend. In doing this, they tamper with the process, which simply introduces new sources of variation. One of the insights of TQM is that managers should avoid such actions. W. Edwards Deming felt that tampering is one of the major failings of modern managers.

For more information, see: W. Edwards Deming, *Out of the Crisis* (Cambridge, MA: MIT Center for Engineering Study, 1986); Brian L. Joiner, *Fourth Generation Management* (New York: McGraw-Hill, 1994); William Lareau, *American Samurai* (New York: Warner Books, 1991).

Expected Product This is a way of describing a customer's *minimum* requirements for benefits from a product or service.

An organization, just to stay in business, must seek to deliver the expected product to customers, but to excel, it must go beyond this to deliver more utility and better problem solutions than expected. In other words, a goal of management must be customer delight. (See also *Customer Delight.*)

For more information, see: Brian L. Joiner, *Fourth Generation Management* (New York: McGraw-Hill, 1994); Tom Peters, *Liberation*

> [Being a statistician] I couldn't understand why people would only want to look at two data points. Finally it became clear to me. With any two data points, it's easy to compute a trend: "Things are down 2% this month from last month. This month is 30% above the same month last year." Unfortunately, we learn nothing of importance by comparing two results when they both come from a stable process . . . and most data of importance to management come from stable processes.
>
> *Brian L. Joiner*

> Take cosmetics. The industry has factory sales of over $3 billion. Yet not a single American woman buys a single penny's worth of cosmetics. Charles Revson, the entrepreneurial genius who built Revlon into the thriving business it is today, has said, "In the factory we make cosmetics. In the store we sell hope."
>
> *Theodore Levitt*

Management (New York: Alfred A. Knopf, 1992); Y.S. Chang, George Labovitz, and Victor Rosansky, *Making Quality Work* (New York: HarperBusiness, 1993); Thomas H. Berry, *Managing the Total Quality Transformation* (New York: McGraw-Hill, 1991); Richard C. Whiteley, *The Customer-Driven Company* (Reading, MA: Addison-Wesley, 1991).

Experimental Design This is another term for *design of experiments* (see this term.)

Explorers This is a metaphor for describing managers who tend to be very open to new ideas for how to attack problems and make positive changes. Explorers generally show leadership behavior as their normal way of operating.

> For more information, see: Craig R. Hickman, *Mind of a Manager, Soul of a Leader* (New York: John Wiley & Sons, 1990).

Extended Enterprise This is a broader definition of the organization as including its suppliers, customers, and other stakeholders, such as the community and larger society. Any definition that does not include the link between all these parties can cause the organization to make poor decisions. While decisions may seem good for the traditionally defined company in the short term, they can have negative consequences for the extended enterprise that will ultimately reflect back on the firm.

The notion of extended enterprise becomes of strategic importance as companies link more closely to their customers through the use of technology and teams and become more dependent on each other as customers narrow the number of suppliers they use. This concept has encouraged companies to connect up in partnerships and brought government agencies and nonprofit organizations together to work on problems of mutual interest. It has led to new processes that begin in one organization and end in another. This idea helps managers appreciate that the welfare of the organization as traditionally defined is connected inextricably with the other members of the extended enterprise.

> For more information, see: Richard J. Schonberger, *Building a Chain of Customers* (New York: The Free Press, 1990); Kenneth

To supply the wants and needs of a consumer, society entrusts wealth-producing resources to the business enterprise.

Peter Drucker

Primozic, Edward Primozic, and Joe Leben, *Strategic Choices: Supremacy, Survival, or Sayonara* (New York: McGraw-Hill, 1991); Tom Peters, *Liberation Management* (New York: Knopf, 1992); Peter M. Senge, *The Fifth Discipline* (New York: Doubleday Currency, 1990).

External Customer This is a person or organization that pays for a company's goods and services and who is not a formal or legal part of the enterprise. This is another term for Big C customer. It contrasts with the idea of internal or little c customer.

For more information, see: Richard C. Whiteley, *The Customer-Driven Company* (Reading, MA: Addison-Wesley, 1991); Richard J. Schonberger, *Building a Chain of Customers* (New York: The Free Press, 1990); Joseph M. Juran, *Juran on Leadership for Quality* (New York: The Free Press, 1989).

External Failure Costs These are costs generated when a defective product or service is delivered to a customer. They include the costs to repair or replace this item or to offer additional services for no charge. (See *Cost of Poor Quality*.)

The expenses to an automobile manufacturer for a recalled vehicle are an example. Warranty costs are another example. We may also view these as the costs of nonconformance to requirements of paying customers. Such costs include legal fees and litigation, scrap, rework, testing, replacement orders, inventory depletion, mistakes, reruns, and so on. You can also include the lost goodwill and future business of customers who have experienced poor quality. There is an informal "rule of ten," which suggests that external failure costs are ten times what it would cost to prevent them in the first place.

For more information, see: Philip B. Crosby, *Quality Is Free* (New York: McGraw-Hill, 1979); Philip B. Crosby, *Quality Without Tears* (New York: McGraw-Hill, 1984).

External Linkages This refers to those linkages that exist between an organization and its suppliers and customers. Upstream linkages are with suppliers and downstream linkages are with channels of distribution or consumers.

As a strategic initiative, external linkages can provide significant competitive advantage and access to markets. Linking up with an oil company to send the oil company's customers an advertisement for your product in the monthly fuel bill envelope is an example of effective use of linkages. It serves as a good substitute for expanding one's own direct sales and dealer network, avoiding internal infrastructure expenses.

For more information, see: Kenneth Primozic, Edward Primozic, and Joe Leben, *Strategic Choices: Supremacy, Survival, or Sayonara* (New York: McGraw-Hill, 1991).

Extrinsic Reward This is any payment, premium, or other recognition given to an individual by another person or by an organization. It includes the traditional rewards decided on by managers for their subordinates. This can include recognition of the performance of a team by another team.

The conventional wisdom is that such rewards are necessary to motivate particular behaviors in people. In general, extrinsic rewards, especially bonuses and salary increases, have only a short-term effect on employee motivation. Further, they direct employees toward doing certain things to gain the rewards, whether their behaviors are in the best interests of customers or not. They also frequently lead to manipulation of the system, such as posting future sales as if they came in during the current period or postponing returns to make current numbers look good in order to receive the reward.

Extrinsic rewards, especially those that serve as tokens of well done work, and given to team members either by top management or, better, by members of other teams, can serve as valuable forms of recognition and feedback. However, they should not serve as the prime motivation of performance or as the goal of behavior, for the reasons just mentioned. At Xerox Corporation, team awards to other teams, for example, are a common phenomenon. Awards for quality performance are given to teams and even whole organizations within IBM. However, these awards provide the positive feedback employees need

Do rewards work? The answer depends on what we mean by "work." Research suggests that, by and large, rewards succeed at securing one thing only: temporary compliance. When it comes to producing lasting change in attitude and behavior, rewards, like punishment, are stikingly ineffective.

Alfie Kohn

to understand how well they are doing and are not viewed as the prime motivators of behavior. We can contrast extrinsic rewards with intrinsic rewards. These have to do with the sense of accomplishment individuals experience from knowing they have done a good job and contributed to their organization or team regardless of whether others have acknowledged these actions.

For more information, see: Jack D. Orsburn, et al., *Self-Directed Work Teams: The New American Challenge* (Burr Ridge, IL: Irwin Professional Publishing, 1990); Alfie Kohn, *Punished by Rewards* (Boston: Houghton Mifflin, 1993).

Facilitator This is a title for anyone who helps a process or project team apply TQM tools and methods. Sometimes the word is used interchangeably with coach or quality advisor. Another definition is a person with the responsibility of helping plan and manage group processes associated with a meeting or project. A facilitator is not necessarily a project team or meeting leader, but someone who helps the team or meeting proceed according to plan.

New process teams often find it difficult to carry out the principles of formal process improvement, using statistical process control methods and data analysis techniques without help. Many organizations have found it effective, therefore, to have facilitators who can guide teams through meetings and can provide advice and counsel as needed during the team's life. Even experienced employees who have worked on multiple process teams often will want help from a facilitator. Frequently these facilitators are other employees who take this on as an extra responsibility. However, they may also be people dedicated full-time within the enterprise to carry out this function.

For more information, see: Charles C. Manz and Henry P. Sims, Jr., *Business Without Bosses* (New York: John Wiley & Sons, 1993); Peter R. Scholtes, *The Team Handbook* (Madison, WI: Joiner Associates, 1988).

> Quality advisors attend team meetings but are neither leaders nor team members. They are "outsiders" to the team in many ways, and can maintain a neutral position. One of their most important jobs arising from this neutrality is to observe the team's progress, evaluating how the team functions, and use these observations to help the team improve its process.
>
> *Peter R. Scholtes*

Facilities Control This is a type of process control that focuses on the maintenance of physical items, such as tools, machinery, instruments, and even buildings.

Factor A statistical term, this is a variable characteristic or condition that influences an effect or an outcome of a single operation or process step.

In design of experiments, engineers often examine various factors to find out if they cause certain effects they are looking for. This is called factorial experimentation, and using DOE techniques, they can test for one or more influencing factors in a single experiment. The purpose of this kind of experiment is to determine if particular changes, for example, to a manufacturing process, improve or hinder the way the process operates. Several statistical techniques exist to undertake and analyze the results of such experiments.

> For more information, see: N. Logothetis, *Managing For Total Quality: From Deming to Taguchi and SPC* (Englewood Cliffs, NJ: Prentice-Hall, 1992); Phillip J. Ross, *Taguchi Techniques for Quality Engineering* (New York: McGraw-Hill, 1988); John L. Hradesky, *Productivity and Quality Improvement* (New York: McGraw-Hill, 1988); William Lareau, *American Samurai* (New York: Warner Books, 1991).

Fail-Safe Technique This is the application of automatic checking devices to capture or prevent the occurrence of common problems in the design or production process, usually having to do with human errors. The Japanese call such fail-safe production methods poka-yoke.

In the United States another way of describing fail-safing is foolproofing. This refers to any action taken in the design of a process to avoid or reduce the number of human-caused errors. The goal in this is always to achieve zero defects. Common process problems caused by human errors usually result from little, poor, or no documentation of process steps, inconsistent process steps, no training of workers in the process, a process too complicated for a worker to understand, or inadequate control limits. By addressing such causes of process problems, process owners can begin to introduce fail-safe techniques.

> For more information, see: Nikkan Kogyo Shimbun Ltd. (eds.), *Poka-Yoke: Improving Product Quality by Preventing Defects* (Cambridge, MA: Productivity Press, 1988); Sarv Singh Soin, *Total Quality Control Essentials* (New York: McGraw-Hill, 1992); J.M. Juran, *Juran on Planning for Quality* (New York: The Free Press, 1988); Bruce Brocka and M. Suzanne Brocka, *Quality Management* (Burr Ridge, IL: Irwin Professional Publishing, 1992).

Failure Mode Effects Analysis (FMEA) Also known as failure mode effects and criticality analysis (FMECA), this is a method for designing in reliability and minimizing the causes of failure in a product. It focuses on analyzing origins of product failure by examining raw materials, components, and assembly processes. The goal is to determine the probability of failure in these items and take preventive action based on this analysis.

> The purpose of the FMEA is (1) to analyze the probable causes of product failure, (2) to determine how the problem affects the customer, (3) to identify the probable manufacturing or assembly process responsible, (4) to identify the process control variable to focus on for prevention and detection, and (5) to quantify the effect on the customer.
>
> *H.G. Menon*

This technique was first developed by the U.S. National Aeronautic and Space Administration (NASA). The technique includes listing potential failure types for each component or subassembly, then numerically ranking them by frequency of occurrence, probability of detection, and importance on a scale of 1 to 5 (1 = Remote, 5 = Very high). These three numbers are then multiplied together to arrive at the *risk priority number (RPN)*. The RPN guides designers of a process, system, or device to focus on the most important components subject to the greatest possible failure. It can save on repair costs while improving the overall quality of a system.

> For more information, see: D.H. Stamatis, *Failure Mode and Effect Analysis: FMEA from Theory to Execution* (Milwaukee, WI: ASQC Quality Press, 1995); H.G. Menon, *TQM in New Product Manufacturing* (New York: McGraw-Hill, 1993); Bruce Brocka and M. Suzanne Brocka, *Quality Management* (Burr Ridge, IL: Irwin Professional Publishing, 1992); Dev G. Raheja, *Assurance Technologies* (New York: McGraw-Hill, 1991).

Failure Rate This is a measure of the frequency of failures over time. Said another way, it defines the probability of a product failing over its expected life. Other similar terms to mean the same thing are "error rates," "defect rates," "discrepancy rate," and "outage rate." (See also *Bathtub Curve*.)

> For more information, see: Armand V. Feigenbaum, *Total Quality Control*, Third Edition, Revised (New York: McGraw-Hill, 1991); Joseph M. Juran and Frank M. Gryna, *Quality Planning and Analysis*, Third Edition (New York: McGraw-Hill, 1993).

Feedback A term from general systems theory generally defined as information about the outputs or behavior of a system that is sent back into the system to affect succeeding outputs.

There are two types of feedback: negative and positive. Negative feedback inhibits a force driving in a certain direction. For example, a company may have a major advertising program planned for a particular product line, and then receive feedback from experiments that the program will not likely make any difference in sales, which inhibits the company from moving forward with this advertising. Positive feedback reinforces the main driving force in a system. In the advertising example, positive feedback would reinforce the company moving ahead with the advertising program.

In TQM, feedback can also be thought of as signal devices that let a person know how a process is operating to help minimize variation and improve quality. Feedback is an important idea in TQM and the management of organizations. By examining the chains of influence and feedback, we can better understand patterns of behavior that either facilitate or hold back individual, team, and organization success.

> For more information, see: Peter M. Senge, *The Fifth Discipline* (New York: Doubleday Currency, 1990); H. James Harrington, *Business Process Improvement* (New York: McGraw-Hill, 1991).

Feedback Loop This is a graphical devices that shows the relationship among the steps in a process and how each influences the other to reinforce or inhibit behavior.

Understanding organizations as systems requires looking at chains of feedback to see how different parts and behaviors of

> Feedback systems are very important. It's clear that if you cannot measure an activity, you cannot improve it. But measurement without feedback is worthless because you have expended the appraisal effort but not provided the individual with an opportunity to improve. Measurement is the lock—feedback is the key. With their interaction, you cannot open the door to improvement.
>
> *H. James Harrington*

the organization influence each other. With this understanding, managers can better plan how to make improvements by making changes in the system that offer the most leverage. They help managers better deal with causes rather than symptoms. Those who study systems have identified a number of behavior patterns, which they call system archetypes. Understanding these archetypes, which are chains of feedback, allows a manager to intervene in a system at the point of greatest leverage to make improvements. Figure 39 illustrates a simple feedback loop showing how parts of system influence each other. This is a reinforcing loop.

FIGURE 39

A simple feedback loop showing how a commitment to JIT gets reinforced in a company. The curved lines indicate the direction of feedback that influences another person, group, or process.

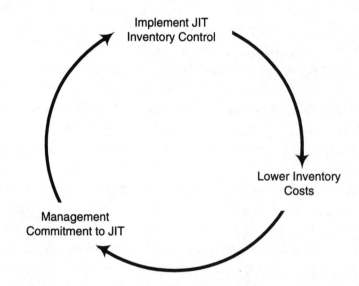

For more information, see: Peter M. Senge, *The Fifth Discipline* (New York: Doubleday Currency, 1990); Peter M. Senge et al., *The Fifth Discipline Fieldbook* (New York: Doubleday Currency, 1994).

Fire Fighting This is an activity performed to eliminate a sporadic problem or quality issue in order to restore the status quo.

Fire fighting is fixing or containing the immediate consequences of a problem right now, often without benefit of solid information about the cause of the problem. It is often described as dealing with symptoms rather than causes and with the idea of crisis management. TQM, with its focus on prevention of defects, identification of problem causes and process variation, and continuous improvement of quality, helps managers minimize the need to do fire fighting and crisis management, activities that add costs but no value to customers.

> For more information, see: William Lareau, *American Samurai* (New York: Warner Books, 1991).

Fishbone Diagram This is also known as a cause-and-effect diagram and is a graphic device that illustrates and categorizes causes and subcauses of a problem. It gets its name from the fact that it takes the form of a fish skeleton, with a long backbone and various ribs jutting off of it. (See *Cause-and-Effect Analysis*, p. 57 for an illustration.)

> For more information, see: Michael Brassard and Diane Ritter, *The Memory Jogger II* (Methuen, MA: GOAL/QPC, 1994); James H. Saylor, *TQM Field Manual* (New York: McGraw-Hill, 1992); or any of several books that cover the tools of TQM.

Fitness for Use This is a phrase used to indicate that a service or product meets a customer's requirements and specifications for its intended use.

The phrase was originally coined by Joseph M. Juran, who made it the central concept of his view of quality. To him, this means a product or service meets customer needs to solve a problem or achieve some goal and that is free of deficiencies and defects. For a company to remain competitive, it has to pay close attention to what it is that makes its outputs fit for use. Then, using quality management tools and techniques, it can make improvements by enhancing a product's capabilities, reducing its deficiencies, and increasing its value by lowering costs.

> For more information, see: Joseph M. Juran, *Juran on Planning for Quality* (New York: The Free Press, 1988); Joseph M. Juran, *Juran on Leadership for Quality* (New York: The Free Press, 1989).

Flowchart This kind of drawing visually defines the steps of a process. Flowcharts provide a picture of how work gets done

> This strategy for allocating attention to issues is nothing more than management by fire control; you wait until a fire breaks out and then you take action, *if* it's sufficiently irritating. A major drawback with this approach, aside from issues of tampering, is that only a handful of the organization's thousands of processes get any attention.
>
> *William Lareau*

and materials move through in a process. It is used to analyze process steps, problem areas, redundant and unnecessary steps, non-value-added activities, and identify areas for improvement.

This is one of the most widely used quality tools because it illustrates graphically all the steps in a process in clear and simple terms. The technique was borrowed from software programming and engineering. See figure 12 under *Block Diagram* (p. 47), or figures 33 and 34 under *Detailed Process Diagram* (p. 129)(other terms for flowchart) for illustrations of flowcharts.

For more information, see: H. James Harrington, *Business Process Improvement* (New York: McGraw-Hill, 1991); Peter R. Scholtes, *The Team Handbook* (Madison, WI: Joiner Associates, 1988); or several other books that include coverage of the tools of TQM.

FMEA This is the acronym for failure mode effects analysis (see this term for definition).

FMECA This is the acronym for failure mode effects and criticality analysis.

Focused Factory This is a factory that manufactures a narrow range of products as a strategy for delivering low cost and high productivity. It grew out of a corporate strategy of "stick to your knitting," that is, attend to those products and services that a company makes best and that customers most value. Then get better and better at manufacturing and delivering those to customers.

A problem with a focused factory is that it can cause companies to seek to reduce costs by eliminating flexibility. This compromises their ability to deliver a variety of product versions to satisfy different market segments. A whole new movement in manufacturing, "mass customization," is now gaining converts. This movement seeks to use technology to gain the benefits of the focused factory, while retaining flexibility to deliver many versions of a product.

For more information, see: George Stalk, Jr., and Thomas M. Hout, *Competing Against Time* (New York: The Free Press, 1990);

> Constructing flowcharts disciplines our thinking. Comparing a flowchart to the actual process activities will highlight the areas in which rules or policies are unclear or are even being violated. Differences between the way an activity is supposed to be conducted and the way it is actually conducted will emerge.
>
> *H. James Harrington*

Peter B.B. Turney, *Common Cents: The ABC Performance Breakthrough* (Portland, OR: Cost Technology, 1991); B. Joseph Pine II, *Mass Customization: The New Frontier in Business Competition* (Cambridge, MA: Harvard Business School Press, 1993).

Focus Group This is a group of customers brought together in a meeting (or a series of meetings) to learn what they think and feel about a service, product, or process. A focus group can also help generate ideas on how to improve goods or services.

Focus groups provide a method any organization can use to gather information about customer needs, wants, and expectations for the benefits products and services a company now offers or might offer in the future. Advertising companies often use focus groups to test commercials and get ideas for improvement. Political candidates use this technique to refine their message and learn what voters think. This is also a technique for better understanding what is going on inside a company by bringing employees together to discuss the company, its culture and direction, and how employees feel about any variety of organizational issues.

A focus group usually consists of six to ten people sitting around a table. A person skilled in directing discussion and eliciting participation leads the group. Before hand, this individual develops a set of questions to cover in the session, often with time parameters to keep the discussion on track. Market research firms are often hired to conduct focus groups. They hold these sessions in special rooms with two-way mirrors so the client can observe the interchange. A focus group session is always audio- or videotaped for further review and analysis. Afterward, the person or company conducting the session prepares a detailed report summarizing responses. Focus groups can generate good ideas for improvement and provide insight into what people are thinking, but they are only one tool for better understanding customers and their needs.

For more information, see: Richard C. Whiteley, *The Customer-Driven Company* (Reading, MA: Addison-Wesley, 1991); David L.

> The focus group is probably most influential when executives sit behind one-way glass and watch customers complain. It's like the ghost of Christmas Present showing Bob Cratchit's house to Ebenezer Scrooge. Nothing else can change an executive's attitude so dramatically.
>
> *Richard C. Whiteley*

Morgan, *Focus on Groups as Qualitative Research* (Newbury Park, CA: Sage Publications, 1988).

Focus Setting This is a method teams can use to concentrate attention on a specific outcome or mission. It incorporates these steps:

- *Define a team mission* (for example, provide a particular service on schedule).
- *Determine improvement opportunities* (identify specific parts of a process or product that need improvement).
- *Select an improvement opportunity to work on.*
- *Establish specific quantitative goals for accomplishing the mission.*
- *Use TQM improvement tools, including the PDCA cycle to achieve goals.*

For more information, see: James H. Saylor, *TQM Field Manual* (New York: McGraw-Hill, 1992).

Foolproofing This is the act of building safeguards into any process or product to prevent or reduce the occurrence of human-caused errors.

This is accomplished using a variety of techniques. One of the most common with products is redundancy of parts; another involves putting locks or blocking access to functions; in software it could be denying a user access to the actual software code, merely access to its use.

Juran gives the following categories for classifying foolproofing methods:

- *Elimination.* Suggests modifications in technology to eliminate operations that are prone to error.
- *Replacement.* Retains the error-prone method but replaces human beings with machines such as robots.
- *Facilitation.* Human operators are given the means to minimize possible errors, such as color coding of parts.

- *Detection.* This method is less about preventing errors than finding them at the earliest point in the process.

- *Mitigation.* Here the process includes means to avoid damage that might come from a human error, such as including fuses in an electrical system.

For more information, see: Sarv Singh Soin, *Total Quality Control Essentials* (New York: McGraw-Hill, 1992); Joseph M. Juran, *Juran on Planning for Quality* (New York: The Free Press, 1988).

Force Field Analysis This is a technique used to identify forces that are for or against a particular action or change and how much force or influence each may exert in facilitating or restraining any change initiative.

Force field analysis is a modification of the pro and con list method for making or not making a change. The difference is that here you do not focus so much on reasons as on forces for and against. This analysis gives you information for (1) deciding whether, under the circumstances, you should make the change; and (2) understanding and dealing with the forces against to ease the change into place. In undertaking the analysis, you try to quantify the forces for and against change, giving each a weight, so as to better understand how they balance each other. In assigning weights, you should research each force so the weight of each is somewhat objective. The length of the arrows on the force field diagram indicate the weights given to each force. Facilitating forces include those that ease the change into happening. Restraining forces are those that bring about significant (and seemingly) negative consequences when the change occurs. Using force field analysis, managers can work at dealing with those factors that may restrain change and movement toward some improvement. Figure 40 illustrates a simple force field diagram for a retail business to understand forces for and against expanding to a new location.

FIGURE 40

A force field analysis diagram for thinking about opening a retail business in a new location. This analysis indicates that this would be a good idea.

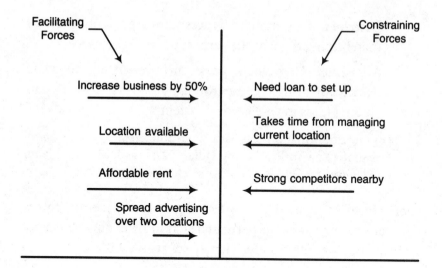

For more information, see: N. Logothetis, *Managing For Total Quality: From Deming to Taguchi and SPC* (Englewood Cliffs, NJ: Prentice Hall, 1992); James H. Saylor, *TQM Field Manual* (New York: McGraw-Hill, 1992); Bruce Brocka and M. Suzanne Brocka, *Quality Management* (Burr Ridge, IL: Irwin Professional Publishing, 1992); Richard C. Whiteley, *The Customer Driven-Company* (Reading, MA: Addison-Wesley, 1991).

14 Points These are W. Edwards Deming's management practices designed to provide the basis for a management system that focuses on improving quality and productivity in any kind of organization or business. In considering the 14 Points, it is useful to remember that Deming has logically derived these from his understanding of systems and statistical variation. These two ideas suggest a holistic view of organization function and the fact that the multiple components in organization processes will always have some variation in them that managers can chart, understand, and decrease. Starting from these assumptions, managers can better see the logic that underpins these points.

Deming's 14 Points for Management

1. Create constancy of purpose to improve products and services.
2. Adopt the new philosophy of quality improvement.

3. Cease dependence on inspection as the way to achieve quality.

4. Stop making purchase decisions just on price alone, work instead with fewer suppliers.

5. Improve constantly and forever processes for planning, manufacture, and service.

6. Institute training on the job.

7. Practice leadership.

8. Drive fear from the workplace.

9. Break down barriers between staffs.

10. Eliminate slogans, targets, and exhortations for workers.

11. Get rid of numerical quotas for workers and numerical goals for managers.

12. Remove obstacles that rob workers of pride of work and eliminate annual merit or rating systems.

13. Educate employees and implement self-improvement for all.

14. Put everyone to work on transforming the enterprise.

It is useful to note that some companies have made Deming's operating principles their entire management philosophy and have prospered because of it. One example is the Zytec Corporation in Minneapolis, which won a Baldrige Award.

For more information, see: W. Edwards Deming, *Out of the Crisis* (Cambridge, MA: MIT Center for Advanced Engineering Study, 1986); Mary Walton, *The Deming Management Method* (New York: Perigee, 1986).

Frequency Distribution This term describes the grouping of data into classes and shows the number of observations for each class.

For example, if you are measuring the time it takes to complete one step in a process, you can take 100 observations, and then group time differences by 10-second intervals. This might also show the frequency distribution of measurement intervals of .1 inch for 150 machined shafts taken as samples from a lot. The frequency distribution would then show how many observations fall into each interval or measurement. We usually display this information in a frequency histogram (see figure 41).

> For more information, see: N. Logothetis, *Managing For Total Quality: From Deming to Taguchi and SPC* (Englewood Cliffs, NJ: Prentice-Hall, 1992); Hy Pitt, *SPC for the Rest of Us* (Reading, MA: Addison-Wesley, 1994); George L. Miller and LaRue L. Krumm, *The Whats, Whys & Hows of Quality Improvement* (Milwaukee, WI: ASQC Quality Press, 1992).

Frequency Histogram This a graph showing the frequency distribution of observations of any type of data. It is usually displayed as a bar chart. Figure 41 shows a typical frequency histogram for different lengths of steel shafts.

FIGURE 41

A frequency histogram showing the quantities of different lengths of 150 steel shafts from a manufacturing process.

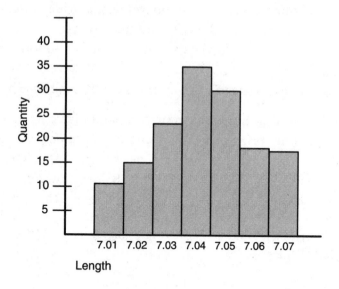

> For more information, see: Hy Pitt, *SPC for the Rest of Us* (Reading, MA: Addison-Wesley, 1994); George L. Miller and LaRue L. Krumm, *The Whats, Whys & Hows of Quality Improvement* (Milwaukee, WI: ASQC Quality Press, 1992).

Functional Decomposition This refers to the breaking apart of functional departments into their discrete activities. This is also a method for identifying all activities of departments.

Managers do functional decomposition when using ABC accounting because ABC attaches cost measurements to activities, not to departments. Variations of the assignment can be to assign costs to processes or cost objects. To perform this process, managers must identify all the activities to which they are prepared to assign costs. Figure 42 illustrates an example of decomposing the activities of a marketing department.

> For more information, see: Peter B.B. Turney, *Common Cents: The ABC Performance Breakthrough* (Portland, OR: Cost Technology, 1991).

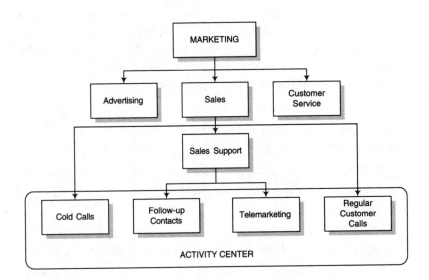

FIGURE 42

An example of functional decomposition to identify an activity center.

Functional Organization This refers to a particular way of structuring an organization around functional areas such as accounting, distribution, legal, product development, marketing, and manufacturing. This organizational structure is becoming outmoded as companies move toward structures that facilitate cross-functional cooperation to efficiently deliver outputs to customers.

The new approach calls for identifying processes and assigning them to various teams, each with its own process owner. This often cuts across old functional areas. For example, a product design process team might include engineers, designers, manufacturing employees, and marketers. All of these functional areas are part of the product development process, and it makes sense for them to work together across functional lines. The goal in doing this is to maximize utility for the customer while controlling costs and identifying potential problems early and taking preventive action. All this does not mean that traditional functional structures are going away. However, it does suggest that both across functions and within functional areas, the focus is on how different areas affect each other and how they might cooperate to prevent problems, keep costs down, and improve quality.

> For more information, see: Kenneth Primozic, Edward Primozic and Joe Leben, *Strategic Choices: Supremacy, Survival, or Sayonara* (New York: McGraw-Hill, 1991); James Brian Quinn, *Intelligent Enterprise: A Knowledge and Service Based Paradigm for Industry* (New York: The Free Press, 1992); Charles Handy, *The Age of Unreason* (Boston: Harvard Business School Press, 1989).

Functional Silos Also called chimneys of excellence, within an organization these are inwardly focused departments that become very efficient at executing their activities, even when this is at the expense of efficiency in other parts of the organization. They are more concerned with their own issues than those of the organization as a whole.

Consider a hotel: Each worker you connect with—at the bellstand or front desk, from room service or housekeeping—is normally part of a different, "vertical" (functional) department. Yet your experience of the hotel is completely "horizontal." You experience "Hyatt." You don't think of Hyatt's room service *department*, etc. So it is, too, in your dealings with IBM, Buckman Labs or MCI.

Tom Peters

As companies move to ABC accounting systems or to cross-functional teams and process management, functional silos become impediments to agility and effectiveness. These departments tend to optimize their performances often at the expense of other departments. This can cause various functional areas to work at cross purposes with each other, which results in the company as a whole suboptimizing its performance. It was a common problem, for example, within large U.S. automotive companies in the 1970s and 1980s when multiple divisions produced cars that essentially were very similar, if not the same. As organizations move toward process management, the ability to move across functional organizations becomes more critical, and along with it, as a natural consequence, the elimination of functional silos, which no longer add value to organizational processes.

For more information, see: Tom Peters, *Liberation Management: Necessary Disorganization for the Nanosecond Nineties* (New York: Alfred A. Knopf, 1992); Rosabeth Moss Kanter, Barry A. Stein, and Todd D. Jick, *The Challenge of Organizational Change* (New York: The Free Press, 1992).

Functional Team This is a team made up of representatives from only one functional area, for example, only people from sales or accounting.

The idea is that these groups of employees work as teams, with the goal of improving existing work processes, applying statistical process control tools and techniques.

For more information, see: Jack D. Orsburn et al., *Self-Directed Work Teams: The New American Challenge* (Burr Ridge, IL: Irwin Professional Publishing, 1990); Peter R. Scholtes, *The Team Handbook* (Madison, WI: Joiner Associates, 1988).

Funnel Experiment This is a demonstration that dramatizes the effects of *tampering* (see this term). Often performed using a device called a quincunx, marbles are dropped through a funnel and fall through a series of nails arranged in groups of five,

> If anyone adjusts a stable process to try to compensate for a result that is undesirable, or for a result that is extra good, the output that follows will be worse than if he had left the process alone.
>
> *W. Edwards Deming*

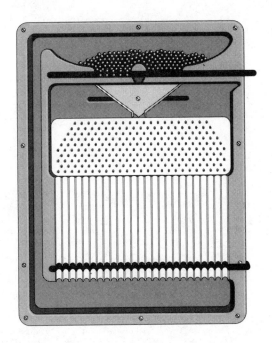

FIGURE 43

A quincunx used to demonstrate the funnel experiment.

then land on a flat-surfaced target area below. This activity illustrates that adjusting a stable process to compensate for an undesirable or a lucky result will generate an outcome worse than if no adjustment had been made. Figure 43 illustrates the device often used in the funnel experiment. Dr. W. Edwards Deming used the funnel experiment to demonstrate a number of concepts that have to do with process control and what happens when you tamper with a process. (See also *Quincunx*).

For more information, see: Brian L. Joiner, *Fourth Generation Management* (New York: McGraw-Hill, Inc., 1994), W. Edwards Deming, *Out of the Crisis* (Cambridge, MA: MIT Center for Advanced Engineering Study, 1986); William W. Scherkenbach, *Deming's Road to Continual Improvement* (Knoxville, TN: SPC Press, 1991).

Gage Repeatability and Reproducibility (GR&R) This is the process for evaluating the accuracy and precision of a gaging instrument by establishing whether measurements taken with the device can be repeated and are reproducible (that is, give nearly the same measurement over and over again). Engineers undertake GR&R studies to establish how much variation results from the measurement process itself, which includes the instrument and measurement techniques.

> For more information, see: Armand V. Feigenbaum, *Total Quality Control*, Third Edition, Revised (New York: McGraw-Hill, 1991); H.G. Menon, *TQM in New Product Manufacturing* (New York: McGraw-Hill, 1992).

Gainsharing This is a reward process by which an organization shares the gains in productivity in an organization or company between owners (for example, stockholders) and employees. To make gainsharing work, Mark Graham Brown, a consultant in quality management, makes the following recommendations:

- Design the plan with employee input—otherwise they won't feel ownership and commitment to it.
- Make sure the size and frequency of payments are often enough and large enough to make a difference to employees. Otherwise they will lose interest.

- Tie payments to individual and team performance. This allows the company to recognize the extraordinary efforts of individuals as well as the effort of the entire team.
- Make sure the performance measures are those that employees can influence or control. If employees, by their efforts, cannot control performance improvements, then the plan will not motivate improvements.

Gainsharing approaches to employee compensation and rewards have become popular in organizations that want to generate greater employee involvement in decision making, skills development, and improvements in productivity. The potential problem with gainsharing, like any kind of bonus program, is that it gets people focused on doing what will make them look good rather than on what's good for the entire company. That's why it must be very carefully administered. Another problem is that unions may oppose it as a way to control pay by management. There is no simple answer to the best approach for compensation and different firms continue to experiment.

> For more information, see: Jack D. Orsburn et al., *Self-Directed Work Teams: The New American Challenge* (Burr Ridge, IL: Irwin Professional Publishing, 1990); Stephen George and Arnold Weimerskirch, *Total Quality Management* (New York: John Wiley & Sons, 1994); Mark Graham Brown, "Paying for Quality," *The Journal for Quality and Participation*, September 1992, pp. 38-43; Brian L. Joiner, *Fourth Generation Management* (New York: McGraw-Hill, 1994).

Gantt Chart This is a type of bar chart often used to display planned and completed activities in specific time periods (for example, planned sales next to actual sales by quarter).

This type of chart is most widely used to manage projects because you can display clearly who does what by when. It then becomes a control chart for determining quickly if you are on schedule, behind, or ahead. In project management it also helps a leader document how much time it should take to perform each task in a process. It is a useful tool for determining the crit-

When employees seek individual rewards, they begin to chase the indicator that triggers that reward. But chasing indicators is a major source of problems: all the pieces look good in their own right, but somehow fail to add up to organizational success.

Brian L. Joiner

ical path for completing the process. Figure 44 illustrates a simple Gantt chart for the addition of space in an office building.

NAME	January 1	January 8	January 15
Begin Project 0 days	◆		
Select Architect 1 day	▢		
Draw Up Plans 1 week	▭		
Corporate Signoff 1 day		▢	
Begin Construction 0 days		◆	
Lay Floor 1 day		▢	
Rough in Walls 2 days		▭	
Hang Drywall 2 days		▮	
Paint Walls 1 day		▮	
Paint Trim 1/2 day		▮	
Lay Carpet 1 day		▢	
Install Doors 1/2 day		▮	
Install Workstations 2 days		▭	
Hook up Phones 1 day			▢
Project Finished 0 days			◆

▭ = Critical Process　▮ = Secondary Process　◆ = Milestone

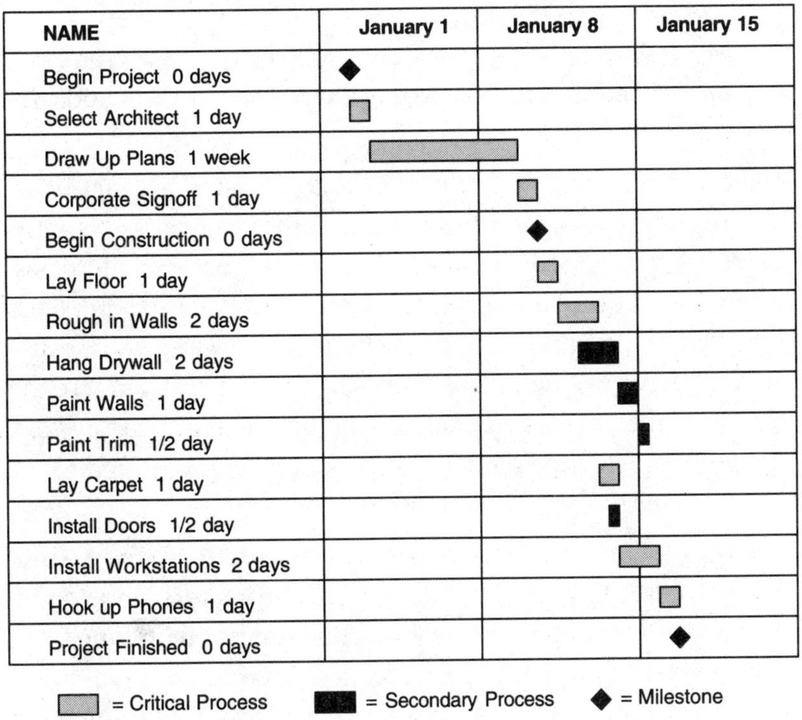

FIGURE 44

An example of a Gantt chart for the addition of new space in an office building.

For more information, see: Harold Kerzner, *Project Management: A Systems Approach to Planning, Scheduling, and Control,* Fourth Edition (New York: Van Nostrand Reinhold, 1992).

Gap This is the change between how things, organizations, or processes are performing today (baseline) and how they should be according to your vision, a benchmark, and/or customer requirements.

Gap analysis has become an increasingly important part of any study of how an organization or process works. In an organizational assessment, gaps are measured by looking at the differences between actual performance and expected performance. Learning the reasons for such gaps involves a thorough analysis of organizational structure, business financial requirements, customer expectations, process performance, information technology implementation, skills of workers, and similar aspects that affect organizational performance. This analysis can then provide guidelines for closing gaps.

> For more information, see: Valarie A. Zeithaml, A. Parasuraman, and Leonard L. Berry, *Delivering Quality Service* (New York: The Free Press, 1990).

Generic Product This term refers to a product class that is undifferentiated in any way from others in the same class. It is a useful concept for discussing the basic benefits customers purchase when they acquire a product. In reality, there are no completely undifferentiated products. Even commodities can be differentiated in terms of grades and other characteristics. Compare this idea with the *augmented product concept* (see this term).

> For more information, see: Theodore Levitt, *Marketing for Business Growth* (New York: McGraw-Hill, 1974).

George M. Low Trophy This prestigious award was given annually by NASA to those contractors, subcontractors, and suppliers of the space agency that had consistently maintained and improved their quality performance. It was also previously called the NASA Excellence Award for Quality and Productivity. Large and small businesses compete in one or the other size category. The award is named after a NASA administrator. NASA ended this program in 1993.

> For more information, contact: American Society for Quality Control, 611 E. Wisconsin Avenue, P.O. Box 3005, Milwaukee, WI 53201-3005; telephone (800) 248-1946.

Goal There are many definitions of this term; however, they all involve a statement of a desired level of performance to be attained. A goal leads to actions directed toward its achievement. Goals are often expressed as measurable targets. The term *goal* is used interchangeably with the term *objective*.

In process work we can define goals as customer expectations and requirements that a company must strive to meet. Goal setting has received much attention in recent years as pro-

---

For years, almost instinctively, consumer goods manufacturers have recognized that the generic product is perhaps the least important part of the product itself. In clothing, it is not dresses one sells, but fashion. In cigarettes, the point reaches its most extraordinary extreme. Only a particularly prudish observer can fail to come to any conclusion other than American cigarette companies literally sell only one product—sex.

Theodore Levitt

Not every end is the goal. The end of a melody is not its goal, and yet if a melody has not reached its end, it has not reached its goal. A parable.

Friedrich Nietzche

cedures for articulating them have improved. For example, many companies are using benchmarking to discover what is possible. They then use the best practices of other companies as goals for themselves and set about reengineering their processes to achieve them. As opposed to an organization's mission or vision, goals are usually quantifiable targets that it should achieve within a fixed time period of a few months to a year.

Sometimes companies pay too much attention to quantifiable goals or end points and very little attention to the processes for achieving them. This diverts them from caring or learning about how to improve their processes. In those companies, we often see people "making their numbers" by manipulating the system so as to look good at the end of the month or year. Such behavior ultimately undermines competitiveness because it is internally focused, based on pleasing the boss, and leaves the company vulnerable to those firms that can deliver higher-quality goods with lower prices.

> For more information, see: David A. Garvin, *Managing Quality: The Strategic and Competitive Edge* (New York: The Free Press, 1988); James H. Saylor, *TQM Field Manual* (New York: McGraw-Hill, 1992); Brian L. Joiner, *Fourth Generation Management* (New York: McGraw-Hill, 1994); Joseph M. Juran and Frank M. Gryna, *Quality Planning and Analysis*, Third Edition (New York: McGraw-Hill, 1993).

Go/No-Go This describes the condition of a unit or product in which one of two circumstances prevail: go (conforms to requirements or specifications) or no-go (does not conform).

One advantage of this approach is that it can require less time and skill to perform than traditional measurement schemes. However, an operator needs to understand the requirements a product or component must meet to be acceptable. That often means measurements still have to be taken. The most frequent use of go/no go decisions is in the use of components in manufacturing. Companies also use this approach in deciding to move ahead with a product or project and in process

Whatsoever thou takest in hand, remember the end, and thou shalt never do amiss.

Apocrypha, Ecclesiasticus 7:36

experimentation depending on whether it meets certain hurdles or not.

For more information, see: Ellis R. Ott and Edward G. Schilling, *Process Quality Control,* Second Edition (New York: McGraw-Hill, 1990).

GR&R This is the acronym for gage repeatability and reproducibility.

Guideline This term refers to specific suggested practices to encourage conformance to predetermined standards that are *not* mandatory, just desirable under normal circumstances.

The concept of guideline is an old one, but organizations are using it in fresh ways. For example, guidelines are often the by-product of benchmarking activities or surveys of customer requirements, and experiences drawn from process improvement or reengineering. They help make sure that processes are executed properly and the outputs meet the needs of customers, both internal and external. However, there is also enough flexibility built into a guideline to allow for contingencies. As authority is delegated further down the organization's hierarchy (or the hierarchy is eliminated), the more important guidelines become as a way of helping people make decisions. The one problem with guidelines is that they can quickly evolve into rigid rules and even acquire absolute measures of their execution. This problem is frequently evident in organizations that do not practice process management.

For more information, see: James H. Saylor, *TQM Field Manual* (New York: McGraw-Hill, 1992).

Hierarchical Nature of a Process This refers to the various levels of any process. It is often depicted as a high-level flowchart of the major parts of a process followed by other charts illustrating the subparts of the high-level phases.

For example, if you were to diagram the process by which you get up in the morning and go to work, at a high level you probably would have a chart showing waking up, getting dressed, having breakfast, and going to your car. That would be a high-level process chart. Next lower in the hierarchy would be flowcharts describing the various steps within each of the major ones. For example, the getting dressed process might include documentation on showering, brushing one's teeth, and so forth. A third level down might include documentation on what you do while showering: washing your hair, bathing your body with a soap different than shampoo, and drying off with a towel.

This approach to defining the steps within a process is particularly useful for complex processes where you must identify many subprocesses that contribute to execution of the organization's overall objectives. Sometimes these are known as macroprocesses and micro-processes. Looking at processes like this is a good way to identify redundant steps and eliminate them. (See also *Macro-Process.*)

You cannot "fix" [or improve] macro-processes. Their outputs can be measured, but the causes of problems are hidden deep within a maze of micro-processes. To fix a macro-process, you must identify and understand all of the inputs, events, and outputs of the constituent micro-processes.

William Lareau

For more information, see: H. James Harrington, *Business Process Improvement* (New York, McGraw-Hill, 1991); William Lareau, *American Samurai* (New York: Warner Books, 1991).

Histogram This is a graphical representation of a distributed set of measurement data, usually by frequency of occurrence. Typically this is a bar graph.

A histogram is one of the basic seven tools used in quality performance measurement. Its value lies in the ability of a chart to show quickly, at a high level, variation in a set of data. See *Frequency Histogram* (p. 169) for an illustration.

For more information, see: Michael Brassard and Diane Ritter, *The Memory Jogger II* (Methuen, MA: GOAL/QPC, 1994); Hy Pitt, *TQM for the Rest of Us* (Reading, MA: Addison-Wesley, 1994); and any of several books that cover TQM tools.

Hoshin Planning Also called Hoshin Kanri, this is an approach to planning in which company-wide long-range objectives are set, taking into account the company's vision, its long-term plan, the needs of customers, the competitive and economic situation, and previous results.

> The [hoshin] process ensures three things: First, that the plan can be achieved because the next level has committed to it. Second, that there is a hierarchy of objectives and strategies. Third, and most important, that the plan is reviewed regularly and corrections made to get the entity or ship back on course.
>
> *Sarv Singh Soin*

It is an approach that incorporates managers and employees at all levels of the organization. They work together in a hierarchical fashion, formulating and then specifically implementing these plans by aligning the objectives of each department and employee with those set for the entire company. It originated in Japan. Figure 45 graphically displays Hoshin planning. In this figure, you can see that the process brings in everyone at each layer of the organization and that the process coincides with the PDCA cycle for continuous improvement. The process coordinates every part and level of the company and includes feedback that allows for the adjustment of plans on a company-wide basis.

Hoshin planning may be contrasted with the American-originated Management by Objectives, which focuses on individual objectives. With MBO, individuals are rewarded for achieving personal goals, which can be at odds with those of the

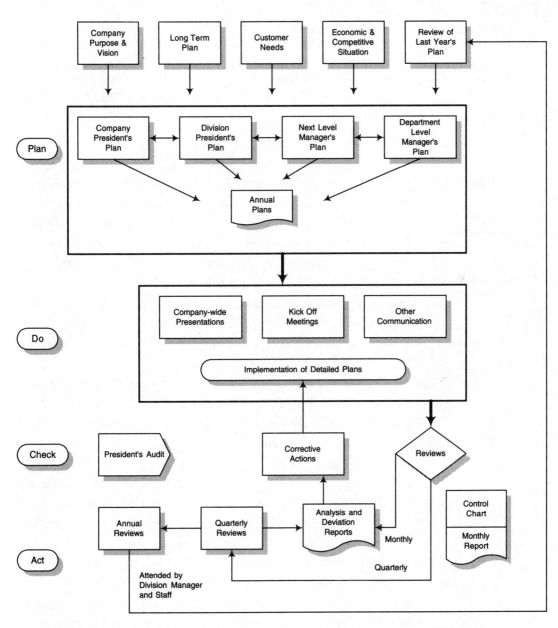

FIGURE 45

Flowchart of Hoshin
planning process.
Adapted from Sarv Singh
Soin, *Total Quality Con-
trol Essentials* (New York;
McGraw-Hill, 1992).

entire organization. The most important aspect of Hoshin planning is the process of carefully aligning the work performed at
each level of the organization with the organization's overall
long-term objectives. (See also *Policy Deployment.*)

For more information, see: Sarv Singh Soin, *Total Quality Control Essentials* (New York: McGraw-Hill, 1992); Bruce Brocka and M. Suzanne Brocka, *Quality Management* (Burr Ridge, IL: Irwin Professional Publishing, 1992); Yoji Akao (ed.), *Hoshin Kanri: Policy Deployment for Successful TQM* (Cambridge, MA: Productivity Press, 1991).

House of Quality This is a graphical device used mainly in the process of designing new products and is part of the process known as *Quality Function Deployment* (QFD). This tool helps managers, teams, or companies correlate customer requirements with product characteristics and to see the strong and weak relationships between the two. It also incorporates an analysis of competitive products in relationship to the new product the company is planning. As a part of the quality function deployment process, the use of this type of diagram makes it possible to bring together several factors into a single figure. The overall goal of this is to translate the voice of the customer into detailed technical requirements.

On the left side we find a list of customer requirements; on the top is a list of potential product characteristics. To the right is a competitive analysis with ratings of various products in terms of customer requirements. In the middle is a matrix that displays a matching of customer requirements with product characteristics. The goal with the use of this chart is to facilitate understanding the relationship between customer needs and product features early in the planning phase of the actual development process. When this is done well, it minimizes rework and helps the company concentrate on those aspects of the product likely to yield the greatest success in the marketplace. This approach, for example, will allow a team to answer the question, what is more important to our customers, a car door that sounds solid when it closes or a body frame that requires a wider door? Figure 46 illustrates a template for building a house of quality diagram.

For more information, see: James L. Bossert, *Quality Function Deployment: A Practitioner's Approach* (Milwaukee, WI: ASQC Quality Press, 1991); N. Logothetis, *Managing For Total Quality: From*

> By recognizing the interrelationships between the engineering properties of the product and the customer's requirements, appropriate actions can be taken at every stage of the product's development, so that the customer needs are anticipated, prioritized, and effectively incorporated into the product.
>
> *N. Logothetis*

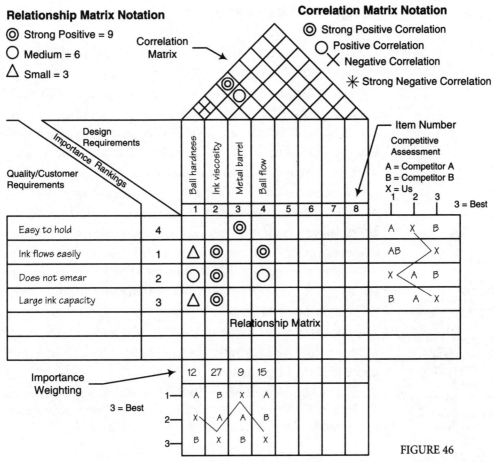

Relationship Matrix Notation

◎ Strong Positive = 9

○ Medium = 6

△ Small = 3

Correlation Matrix Notation

◎ Strong Positive Correlation

○ Positive Correlation

✕ Negative Correlation

✳ Strong Negative Correlation

FIGURE 46

A simple version of the house of quality employed in quality function deployment. In this example, the product is a ball point pen and the various customer requirements are correlated with design and technical requirements. It also shows where the company stands in comparison to major competitors. It looks complicated, but it captures lots of information on one diagram.

Deming to Taguchi and SPC (Englewood Cliffs, NJ: Prentice-Hall, 1992); Sarv Singh Soin, *Total Quality Control Essentials* (New York: McGraw-Hill, 1992); Bruce Brocka and M. Suzanne Brocka, *Quality Management* (Burr Ridge, IL: Irwin Professional Publishing, 1992).

Hypothesis This is an assertion made about some parameter or circumstance, usually used in scientific analysis. With TQM's application of the tools of science and statistics, we see more of the formal use of hypotheses to better understand and deal with business problems and process and quality improvement. This represents a more informed approach to decision making based on data and facts.

Hypothesis consulting is a widely used form of research by management consultants as well. This approach involves defining several hypotheses that may explain the course of events in a process or business. This is followed by experiments and research activities designed to validate or disprove these hypotheses. During the process, managers gather facts, identify findings from those facts, and draw conclusions. From the conclusions come recommendations for action. The beginning of this type of investigation process is the hypothesis, a kind of informed guess, subject to further testing. Originally used in scientific and engineering applications, it is just as applicable to business conditions. For example, a hypothesis might be "customers are willing to pay 5 percent more for a better quality product" or "we can cut costs by 20 percent by outsourcing all data processing." The company can then test these hypotheses by various experiments. Figure 47 illustrates the use of hypotheses to use data to come to sound conclusions.

For more information, see: Paul F. Wilson, Larry D. Dell, and Gaylord F. Anderson, *Root Cause Analysis: A Tool for Total Quality Management* (Milwaukee, WI: ASQC Quality Press, 1993).

FIGURE 47

Moving to recommendations using a hypothesis statement. This statement leads to a collection of facts from which findings are made and conclusions drawn that serve as a basis for recommendations.

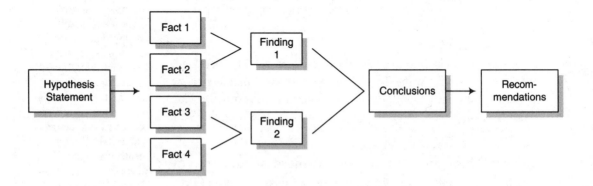

Imperfection This is a departure from customer requirements or standards for the quality of an output. It is another word for defect, blemish, or nonconformity.

Improvement In quality management, this word suggests a decline in errors or enhancement of performance on a continuous or incremental basis.

Improvement of the process includes better allocation of human effort. It includes selection of people, their placement, their training, to give everyone, including production workers, a chance to advance their learning and to contribute the best of their talents. It means removal of barriers to pride of workmanship both for production workers and for management and engineers.

W. Edwards Deming

A formal tenet of TQM is that a company has the obligation to continuously improve the processes by which it delivers its products and services as well as the quality (that is, the value and utility) of its products and services themselves. But beyond the idea of obligation, it is also vital to recognize, as TQM does, that the processes of any organization are always changing in one of two ways: They are either deteriorating, or they are improving. Wear and tear on machines and facilities, laxity on the part of employees, and many other factors operate to introduce variation in the outcomes of any process. Unchecked, this variation increases. TQM includes a variety of tools for monitoring and understanding the sources of variation and provides strategies for reducing it. In a competitive marketplace, customers are constantly looking for more value and utility from their purchases. For this reason, TQM also reminds managers of the importance of improving product and service quality for their company to remain viable.

We may contrast the notion of improvement on a continuous basis with the newer idea of *reengineering*, which suggests dramatic change and performance improvement due to a new and more efficient process to deliver an output. (See also *Reengineering.*)

> For more information, see: W. Edwards Deming, *Out of the Crisis* (Cambridge, MA: MIT Center for Advanced Engineering Study, 1986); Brian L. Joiner, *Fourth Generation Management* (New York: McGraw-Hill, 1994); Arthur R. Tenner and Irving J. DeToro, *Total Quality Management: Three Steps to Continuous Improvement* (Reading, MA: Addison-Wesley, 1992); Michael Hammer and James Champy, *Reengineering the Corporation* (New York: HarperBusiness, 1993); and many other books that deal specifically with TQM tools and procedures.

Improvement Methodology This refers to a variety of techniques for implementing improvements in organizational processes and outputs.

There are many terms that suggest improvement methodologies: TQM, TQC, continuous improvement, process improvement, and process reengineering are a few. Such approaches are comprehensive, calling for a combination of employee incentives and training, a process view of improvements, and enterprise-wide measurements of performance. These programs may take years of focus to be completely effective, although returns on such investments of time and money often begin quickly—within 18 months—and then actually grow in impact.

> For more information, see: Thomas H. Berry, *Managing the Total Quality Transformation* (New York: McGraw-Hill, 1991); David A. Garvin, *Managing Quality: The Strategic and Competitive Edge* (New York: The Free Press, 1988); Michael Hammer and James Champy, *Reengineering the Corporation* (New York: HarperBusiness, 1993); William Lareau, *American Samurai* (New York: Warner Books, 1991); Arthur R. Tenner and Irving J. DeToro, *Total Quality Management: Three Steps to Continuous Improvement* (Reading, MA: Addison-Wesley, 1992); and many other books that deal with specific TQM tools and procedures.

Inadvertent Errors These are human errors caused by inattention. Such mistakes are unpredictable, always considered unwitting, and usually result from unintentional actions.

If they occur infrequently and result in an output that falls outside control limits, then they are due to special causes and can be addressed quickly. If seemingly inadvertent errors happen frequently in a process, this suggests they are due to common causes that are simply a part of the process itself. This then suggests that either the process needs to be better understood and modified or employees need further training to minimize the possibility of inadvertent errors.

> For more information, see: Joseph M. Juran and Frank M. Gryna, *Quality Planning and Analysis*, Third Edition (New York: McGraw-Hill, 1993); Joseph M. Juran, *Juran on Leadership for Quality* (New York: The Free Press, 1989); William Lareau, *American Samurai* (New York: Warner Books, 1991).

> At the microprocess level, work is often highly repetitive and thereby subject to lapses in human attention, that is, inadvertent errors. A major form of remedy is to try to eliminate the possibility of error, or "foolproof" the process, that is, redesign the process so it is not possible to make the error. The work force is an excellent source of ideas on error proofing.
>
> *Joseph M. Juran*

In-Control Process This refers to any process that, by statistical measurements, is performing within control limits. Another way of stating this is to say the process is in statistical control and that we can attribute all variation in outputs to common causes within the process. Once you have discovered that a process is in statistical control, you can then begin to examine how to reduce the sources of variation in that process and thus in the quality of its outputs.

> For more information, see: W. Edwards Deming, *The New Economics* (Cambridge, MA: MIT Center for Advanced Engineering Study, 1993); Armand V. Feigenbaum, *Total Quality Control*, Third Edition, Revised (New York: McGraw-Hill, 1991); Brian L. Joiner, *Fourth Generation Management* (New York: McGraw-Hill, 1994).

Indicators These are characteristics of products, processes, and services that are measurable and best represent customer satisfaction or quality.

Information System This usually refers to automated (computer-based) systems for collecting, analyzing, storing, and dispensing data and information. It also describes the flow of information that allows one to review, analyze, and take corrective actions within a process or processes.

Recent studies on the role of information and information technology (IT) show that the use of information systems has a major positive effect as an enabler for dramatically improving processes. For one thing, IT makes it possible to gather much information with relative ease about the performance of a process. For another, IT facilitates the redesign of a process to make it faster, easier, more comprehensive, and more effective in responding to customer or market conditions. Another reason the application of well-designed IT improves efficiency is that we can often attribute the redundancy and complexity in any process to uncertainty and the lack of shared information among all parties involved. IT facilitates the rapid gathering and dispersion of information to everyone who needs it in a company as well to suppliers and even to customers.

> For more information, see: Thomas H. Davenport, *Process Innovation: Reengineering Work Through Information Technology* (Boston: Harvard Business School Press, 1993); James W. Cortada, *TQM for Information Systems Management* (New York: McGraw-Hill, 1995); Timothy Braithwaite, *Information Service Excellence Through TQM* (Milwaukee, WI: ASQC Quality Press, 1994).

Inhibitors In TQM, this is a name given to those managers or employees who are reluctant to embrace quality principles and may, in fact, be blocking their implementation through overt or covert actions.

Most people begin as inhibitors because they are not yet persuaded by the philosophy or logic of Total Quality Management and continuous improvement or are simply waiting to see positive results. Inhibitors are usually the last to adopt quality management principles. Companies can minimize the number of inhibitors by embracing cultural values that encourage teamwork, continuous improvement, and an unrelenting focus on customer satisfaction, especially by top management. Without top management "walking the talk," inhibitors will be common in companies because these individuals see that the movement toward TQM is mainly lip service to the latest management fad by those who run the company. (TQM is *not* a fad.)

> For more information, see: William Lareau, *American Samurai* (New York: Warner Books, 1991); Roger Tunks, *Fast Track to Quality* (New York: McGraw-Hill, 1992).

In-Plant Quality Evaluation Program (IQEP) This is a method by which a plant judges how well its controls over product or process quality operate.

Manufacturing plants that are ISO 9000 certified have documented, organized quality evaluation programs that range from the performance of suppliers (for example, of raw materials or components), through its own fabrication processes, all the way out to and including the company's distributors.

> For more information, see: Richard J. Schonberger, *World Class Manufacturing: The Lessons of Simplicity Applied* (New York: The Free Press, 1986); Ronald Cottman, *A Guidebook to ISO 9000 and ANSI/ASQC 90* (Milwaukee, WI: ASQC Quality Press, 1993); John T. Rabbit and Peter A. Bergh, *The ISO 9000 Book*, Second Edition (New York: Amacom, 1994).

Input This is the injection of energy, information, data, labor, or materials into a process required to complete an activity to produce an output or work product. The term was borrowed from information processing in which data are "input" into a computer.

From a systems view, input is one of three generic components. The other two are process and output. All process documentation has as its overarching structure input, process, and output, as shown in figure 48. This world view suggests that several opportunities for process improvement come from focusing on improving the quality of inputs and improving the timeliness of delivery (otherwise known as "just-in-time" inventory management). Many firms now understand the value

As organizations embark on the TQM process, it doesn't take most midlevel managers long to understand and use the jargon of TQM. They talk about support and commitment without any real, internal commitment to Quality. Their words lack substance and are nothing more than glitter. Yet, for TQM to succeed, you have to pay special attention to the middle level management level to build the intrinsic Quality commitment.

Roger Tunks

FIGURE 48

The systems view of a company and the place of inputs. TQM is about managing the whole system.

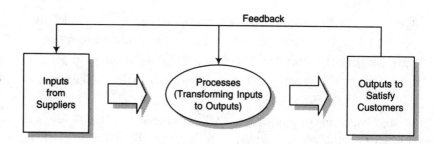

of working more closely with suppliers to ensure that inputs are of the appropriate quality, price, and, as just noted, delivered exactly when needed. Firms that are the suppliers are also developing long-term partnerships with their customers to better serve their needs.

For more information, see: H. James Harrington, *Business Process Improvement* (New York: McGraw-Hill, 1991); Richard J. Schonberger, *Building a Chain of Customers* (New York: The Free Press, 1990); William Lareau, *American Samurai* (New York: Warner Books, 1991); Brian L. Joiner, *Fourth Generation Management* (New York: McGraw-Hill, 1994).

Inspection This includes the acts of measuring, testing, examining, or gaging one or more characteristics of the outputs of a process and then comparing results to specified requirements. The objective is to determine if output features conform to specifications for size, function, appearance, and other characteristics that may be relevant to the product or service.

While there are many tools and techniques for performing inspections, there are two important schools of thought. The more traditional one holds that the company conducts inspections on products or services after the item leaves the production line or the service is performed. The second school of thought—the one promoted by most quality experts—is that inspections should occur all along the process way, from beginning to end. TQM and SPC tools are used to identify problems as early in a process as possible, with the goal of eliminating causes of problems. This approach saves money by eliminating process problems that result in defective outputs at the end of a process. This makes everyone more productive and helps ensure that final outputs will be defect free and meet customer requirements. It is important to appreciate this point. Statistical process control is a way to conduct ongoing inspection of every step in a process as may be appropriate.

> For more information, see: N. Logothetis, *Managing For Total Quality: From Deming to Taguchi and SPC* (Englewood Cliffs, NJ: Prentice Hall, 1992); Armand V. Feigenbaum, *Total Quality Control*, Third Edition, Revised (New York: McGraw-Hill, 1991); Charles A. Mills, *The Quality Audit* (New York: McGraw-Hill, 1989); William Lareau, *American Samurai* (New York: Warner Books, 1991).

> Just as the theme of the traditional inspection activity was "bad parts and products shall not pass," the theme of the new quality approach is "make them right the first time." Emphasis is on defect *prevention* so that routine inspection will not be needed to as large an extent.
>
> *Armand V. Feigenbaum*

Inspection 100% This is the inspection of all units within a lot, sometimes also called screening inspection.

Instant Pudding This is a metaphor employed to suggest the misunderstanding that TQM can be as easily and quickly implemented at a company as it is to make instant pudding.

Coined originally by James Bakken of Ford Motor Company, it implies a practice that W. Edwards Deming was very critical of. He used the phrase to illustrate the fallacy of thinking that quality could be easily improved by some quick fix, such as superficial uses of statistical quality control, glib formulae, or brief use of a consultant. For quality experts such as Deming, the notion of instant pudding is false, yet they find widespread belief in the idea. It is characteristic of the American business culture to want immediate answers. TQM recognizes that quality improvement is an ongoing activity that must be deeply embedded in a company's culture and that quick fixes do not fix anything for long.

> Letters and telephone calls received by this author disclose prevalence of the supposition that one or two consultations with a competent statistician will set the company on the road to quality and productivity—instant pudding.
>
> *W. Edwards Deming*

> For more information, see: W. Edwards Deming, *Out of the Crisis* (Cambridge, MA: MIT Center for Advanced Engineering Study, 1986).

Institutionalize This is the act of making something an integral aspect of an organization's way of life. This action usually affects a cultural value or a process.

Standard techniques to accomplish institutionalization of a value or way of approaching an activity involve changes in measurements, reward and compensation systems, training programs, organizing by process, and leadership espousing new cultural values. For any changes to become institutionalized, they must be adhered to for a period of time, and they must result in improvements in efficiency, productivity, and quality.

For more information, see: James W. Cortada and John A. Woods, *The Quality Yearbook* (New York: McGraw-Hill, published annually); Rosabeth Moss Kanter, Barry A. Stein, and Todd D. Jick, *The Challenge of Organizational Change* (New York: The Free Press, 1992).

Internal Customer Also known as little c customer, this is any person in an organization who receives the output of another fellow-employee's process work.

The thought behind this phrase is that we all treat customers with great respect and politeness, and attempt to satisfy their needs to gain their favor and their business. This same set of behaviors should be applied to fellow employees with whom we work and who are dependent on our work—our internal customers. If we do that, then the work one person does for another is performed to meet the requirements of that person.

For more information, see: Richard C. Whiteley, *The Customer-Driven Company* (Reading, MA: Addison-Wesley, 1991); Kiyoshi Suzaki, *The New Shop Floor Management* (New York: The Free Press, 1993); Christopher W. Hart, "The Power of Internal Guarantees," *Harvard Business Review*, January-February 1995, pp. 64-73, reprint number 95106.

Internal Failure Costs These are the costs caused by defects found in products and services before they reach a customer outside the enterprise. These defects are often discovered by inspection at the end of a process. We can distinguish internal failure costs from failure or defect costs generated by problems once products or services have reached a customer.

The generally acknowledged originator of this concept, Armand V. Feigenbaum, argues that we can compare these types of

> [When hiring], include the attitudes and skills the job itself requires and also the attitudes your organization's vision demands. Even a purchasing clerk or a lathe operator should want to delight customers—both the organization's ultimate customers and the "internal customers" the individual serves.
>
> *Richard C. Whiteley*

costs (a form of the cost of poor quality) by product line and even against competition. He has shown that the cost of quality consists of external failures, internal failures, appraisal costs, and prevention. Workers can study internal failures and use their findings to identify opportunities for improvement. Any lapse in quality in a process that requires rework or repair represents an internal failure cost.

Some examples of internal failure costs include redoing documents because they were improperly typed, overtime caused by poor documentation and execution of processes, engineering changes caused by faulty designs, and so on. A goal of sound quality management is to minimize and even eliminate such costs by continuously reducing the possibility of any process output not meeting specifications.

> For more information, see: Armand V. Feigenbaum, *Total Quality Control*, Third Edition, Revised (New York: McGraw-Hill, 1991); Henry L. Lefevre, *Quality Service Pays: Six Keys to Success!* (Milwaukee, WI: ASQC Quality Press, 1989).

Internal Linkages These are linkages within a value chain. The effect of an order entry process on delivery costs, for instance, is an internal linkage between a primary activity (delivery) and a support function (order entry).

Michael E. Porter defined these linkages as "relationships between the way one value activity is performed and the cost or performance of another." For instance, if a computer manufacturer buys high-quality computer chips, this can speed up the process of manufacturing defect-free microcomputers. Effective linkages between organizations supporting specific processes can enhance competitive advantages for all involved because one can optimize or better coordinate activities that have value in the market.

> For more information, see: Michael E. Porter, *Competitive Advantage: Creating and Sustaining Superior Performance* (New York: The Free Press, 1985).

Intrinsic Reward This usually refers to a feeling of accomplishment or satisfaction within an individual brought about by per-

> Linkages can lead to competitive advantage in two ways: optimization and coordination. Linkages often reflect tradeoffs among activities to achieve the same overall result. For example, a more costly product design, more stringent materials specifications, or greater in-process design, may reduce service costs.
>
> *Michael E. Porter*

sonal satisfaction with performance. The reward is, in effect, given to oneself by mental recognition and appreciation of personal achievement.

W. Edwards Deming linked this notion to his concept of the "joy of work," in which doing well is something pleasing to all people. However, managers do not always understand the implications of this idea. Measurement systems, rewards and incentives, and other actions can actually dampen joy in work. This is because they assume that people work to please others to receive a reward, the dispersion of which someone else controls. If you accept the notion that people can and do enjoy work that they find pleasure in, are successful at, and are complimented and recognized for, then the reward comes from doing the work itself. Management practices that rob employees of the intrinsic rewards of work (and the accompanying motivation) can compromise employee performance.

> For more information, see: W. Edwards Deming, *The New Economics* (Cambridge, MA: MIT Center for Advanced Engineering Study, 1993); Kathleen D. Ryan and Daniel K. Oestreich, *Driving Fear Out of the Workplace* (San Francisco: Jossey-Bass, 1991); James H. Saylor, *TQM Field Manual* (New York: McGraw-Hill, 1992); Alfie Kohn, *No Contest: The Case Against Competition*, Revised Edition (Boston: Houghton Mifflin, 1992); Alfie Kohn, *Punished by Rewards* (Boston: Houghton Mifflin, 1993).

Investment Management In a traditional company, this term means the management of productive capacity to maximize profits from goods and services. In process-focused companies, this refers to the practice of assigning capital investments to areas with the highest promise of returns on investments.

When investing in processes, ABC accounting approaches are frequently used because they document the actual cost of process activities that, when compared to the value of the output of these processes, can lead to a definition of the relative potential yield of an investment in further process improvements.

[Extrinsic rewards] squeeze out from an individual, over his lifetime, his innate intrinsic motivation, self-esteem, dignity. They build into him fear, self-defense, extrinsic motivation. We have been destroying our people, from toddlers through the university, and on the job. We must preserve the power of intrinsic motivation, dignity, cooperation, curiosity, joy in learning, that people are born with.

W. Edwards Deming

For more information, see: Peter B.B. Turney, *Common Cents: The ABC Performance Breakthrough* (Portland, OR: Cost Technology, 1991).

Ishikara Diagram This is a cause-and-effect diagram. (See page 57 for an illustration.)

The diagram allows for the graphical analysis of the causes of effects (or problems) into a variety of categories until the root cause or causes of a particular effect can be identified. This diagram is considered one of the seven basic tools of any improvement effort.

> For more information, see: James H. Saylor, *TQM Field Manual* (New York: McGraw-Hill, 1992); Michael Brassard and Diane Ritter, *The Memory Jogger II* (Methuen, MA: GOAL/QPC, 1994); and many other books on quality improvement tools.

ISO 9000 Standards This is a set of quality standards developed in 1987 by the International Organization for Standardization (ISO). Companies can become ISO certified by undergoing a rigorous audit of their processes.

The purpose of this audit is to demonstrate that companies have documented their processes in 20 different categories and are executing these processes as documented. This certification helps assure other companies that any firm actually does what it says it does, *all the time.* The European Economic Community requires companies that wish to do business with them to have ISO 9000 certification.

There are three major areas of certification:

- **ISO 9001**, which covers all the processes of a company from design and development to procurement, production, testing, installation and service.
- **ISO 9002**, which covers everything except design and development.
- **ISO 9003**, which covers only inspection and testing.

Perhaps a company only seeks certification of one sort, such as ISO 9002, because its customers do not expect it to be involved

ISO 9000 is at the very basic end of the quality evolution and provides you with stability and the minimum attributes for market survival. With ISO 9000 compliance and certification, you will notice increased employee involvement and the ability to more easily correct problems.

John T. Rabbit and Peter A. Bergh

in design and development activities. The up to 20 macro-processes that must be documented (depending on the certification a company seeks) include:

1. Management responsibility
2. Quality system
3. Contract review
4. Design control
5. Documents and data control
6. Purchasing
7. Control of customer-supplied product
8. Product identification and traceability
9. Process control
10. Inspection and testing
11. Control of inspection, measuring, and test equipment
12. Inspections and test status
13. Control of nonconforming product
14. Corrective and preventive action
15. Handling, storage, packing, preservation, and delivery
16. Control of quality records
17. Internal quality audits
18. Training
19. Servicing
20. Statistical techniques

A company need not focus on all these areas, depending on the level of certification sought.

For more information, see: Ronald Cottman, *A Guidebook to ISO 9000 and ANSI/ASQC 90* (Milwaukee, WI: ASQC Quality Press, 1993); Robert Peach, *The ISO 9000 Handbook* (Fairfax, VA: CEEM Information Services, 1993); John T. Rabbit and Peter A. Bergh, *The ISO 9000 Book*, Second Edition (New York: Amacom, 1994).

Joiner Triangle This is a graphical device developed by quality consultant and author Brian L. Joiner to capture what he considers the essence of quality management: the scientific approach, quality, and all one team.

Joiner uses this device to remind managers that they must focus on doing all these things to succeed. The scientific approach includes data collection and analysis techniques using statistical process control. Quality has to do with meeting customer requirements and making customer delight your goal. All one team reminds managers that work gets done by people cooperating with one another as a team with shared goals and understandings. Figure 49 shows the Joiner Triangle.

FIGURE 49

The Joiner Triangle showing the three essential elements of quality management.

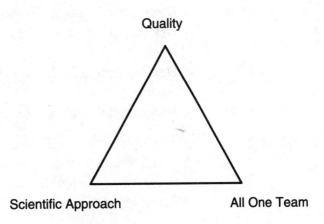

For more information, see: Brian L. Joiner, *Fourth Generation Management* (New York: McGraw-Hill, 1994); Peter R. Scholtes, *The Team Handbook* (Madison, WI: Joiner Associates, 1988).

Joint Planning This is typically defined as planning conducted together by suppliers and customers.

This often results in strategic relationships that benefit both parties and lower costs. For example, the Chrysler Corporation now works closely with its suppliers to design and plan its new products. By giving suppliers the responsibility for coming up with component designs and their improvements over time, a company saves time and money in the execution of its processes and ends up with higher quality final products. In Japan, the notion of *keiretsu* characterizes such long-term relationships between manufacturers and their suppliers. Such relationships benefit all parties.

For more information, see: James F. Cali, *TQM for Purchasing Management* (New York: McGraw-Hill, 1993).

Juran Trilogy® This is a patented set of processes used by Joseph M. Juran to perform (1) quality planning, (2) quality control, and (3) quality improvement through his consulting operations.

As Juran describes these aspects of the Juran Trilogy, he breaks them down in this way:

Quality Planning

- Determine who the customers are.
- Determine the needs of customers.
- Develop product features that respond to customers' needs.
- Develop processes that are able to produce those product features.
- Translate the resulting plans to operating forces.

Quality Control

- Evaluate actual quality performance.
- Compare actual performance to quality goals.
- Act on the difference.

Quality Improvement
- Establish the infrastructure needed to secure annual quality improvement.
- Identify the specific needs for improvement—the improvement *projects.*
- For each project, establish a project team with clear responsibilities for bringing the project to a successful conclusion.
- Provide the resources, motivation, and training needed by the teams to (a) diagnose the causes, (b) stimulate establishment of a remedy, and (c) establish controls to hold the gains.

For more information, see: Joseph M. Juran, *Juran on Leadership for Quality* (New York: The Free Press, 1989).

Just-in-Time (JIT) This refers to inventory control practices, primarily in manufacturing, that call for production of goods to be as close as possible to the time when they are sold. Implied here is the availability of raw materials and components within hours of consumption or provision of a service as needed.

The concept broadened during the 1980s and 1990s to describe strategies that result in a short cycle of manufacturing once a customer has designed or placed an order. Mass customization is a manifestation of JIT manufacturing.

For more information, see: Richard J. Schonberger, *Manufacturing Casebook: Implementing JIT and TQC* (New York: The Free Press, 1987); Richard J. Schonberger, *Building a Chain of Customers* (New York: The Free Press, 1990); William Lareau, *American Samurai* (New York: Warner Books, 1991); B. Joseph Pine II, *Mass Customization* (Cambridge, MA: Harvard Business School Press, 1993); Masaaki Imai, *KAIZEN: The Key to Japan's Competitive Success* (New York: McGraw-Hill, 1986).

Just-in-Time Manufacturing In pure manufacturing terms, this is a materials requirement planning approach in which hardly any inventory of parts or raw materials are kept at the factory and in which little to no incoming inspection occurs of parts or raw materials.

A true JIT system does not use inventory to smooth out discontinuities in output between processes. Instead processes are studied in detail and work is "leveled" so that workers are kept optimally busy while producing only as much output as the next process asks for.

William Lareau

It emerges from a philosophy of never-ending elimination of waste. (One source of wasted costs comes from having larger than needed inventory of parts on hand at any time.) We can contrast the JIT approach with the "Just-in-Case" approach practiced in most companies, where they have invested in large excess inventories of parts they may or may not need, "just in case." W. Edwards Deming noted this, and working closely with suppliers is one of his 14 Points for Management.

Pioneered by Japanese manufacturing companies, and in particular within the automotive industry (especially Toyota), it has now become a widely used set of processes for ensuring low investments in inventory. A critical factor in making this approach work is developing close ties (and even partnerships) with suppliers, as opposed to the more traditional approach of an adversarial relationships with these companies. Doing this involves sharing a factory's information with suppliers as if they were employees of the company.

For example, a tire supplier would have access to an auto plant's production plan for next week to know how many tires to have on-site each day. Paying suppliers by having them draw down their fees electronically from either a line of credit or an account represents another major change in supplier/customer relations that helps make JIT work for both parties. JIT strategies are proving to be cost justified in manufacturing as well as in process and service industries.

For more information, see: Richard J. Schonberger, *Manufacturing Casebook: Implementing JIT and TQC* (New York: The Free Press, 1987); Richard J. Schonberger, *Building a Chain of Customers* (New York: The Free Press, 1990); Michael Hammer and James Champy, *Reengineering the Corporation* (New York, HarperBusiness, 1993); William Lareau, *American Samurai* (New York: Warner Books, 1991); W. Edwards Deming, *Out of the Crisis* (Cambridge, MA: MIT Center for Advanced Engineering Study, 1986); Masaaki Imai, *KAIZEN: The Key to Japan's Competitive Success* (New York: McGraw-Hill, 1986).

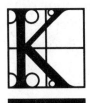

Kaizen This Japanese word is the equivalent of the English phrase *continuous improvement,* in which processes are incrementally enhanced through time to continuously improve overall performance of a process or organization.

The essence of KAIZEN is simple and straightforward. KAIZEN means improvement. Moreover, KAIZEN means ongoing improvement involving everyone, including both managers and workers. The KAIZEN philosophy assumes that our way of life—be it our working life, our social life, or our home life—deserves to be constantly improved.

Masaaki Imai

This approach to management characterizes, perhaps more than any other, the Japanese philosophy of work that is at the heart of their success in delivering high-quality products that have high appeal to customers. Kaizen encourages employees to make suggestions for improvement and often empowers them to implement these improvements. Some Japanese companies receive a hundred or more suggestions for improvement per employee each year.

The underlying assumption is that small improvements, continuously made to a process, will lead to significant positive changes over time. The length of time and the degree of change varies widely. Current research, though, clearly supports the assumption that there is a profound cumulative positive effect on the overall performance of both processes and organizations. Sometimes kaizen does not yield dramatic positive effects. This usually happens when the process needs to be replaced through reengineering.

For more information, see: Masaaki Imai, *KAIZEN: The Key to Japan's Competitive Success* (New York: McGraw-Hill, 1986).

Kanban This is an important tool in the Japanese approach to just-in-time production and inventory control. Kanban literally means card and was developed by Taiichi Ohno of Toyota. The use of kamban works like this: As one group of parts are delivered to the next station in a process, a kanban, or a card, is attached. When this station has used all these parts, it returns the card to its supplier, who uses the card as an order for another group of parts. The use of this tool helps prevent bottlenecks in a process and reduce the costs associated with partially completed products. It is part of a comprehensive approach to production and quality control.

> For more information, see: Masaaki Imai, *KAIZEN: The Key to Japan's Competitive Success* (New York: McGraw-Hill, 1986); William Lareau, *American Samurai* (New York: Warner Books, 1991).

Key Business Processes This refers to the most important processes of a business, usually those that influence how customers view and interact with the company. There are typically less than one dozen of these processes. Often, they are also seen as part of "moments of truth" contacts with customers.

Examples of such key processes are those involved with customer service, maintenance of purchased products, telephone contacts, and customer surveys.

> For more information, see: H. James Harrington, *Business Process Improvement* (New York: McGraw-Hill, 1991); H. James Harrington, *Total Improvement Management* (New York: McGraw-Hill, 1994).

Key Interface This is the primary channel of dialogue between a company and a customer. Managing this interface is crucial to maintaining a long-term positive relationship with customers.

Companies that are among the most admired, such as L.L. Bean, Nordstrom's, AT&T Universal Card, and the Ritz Carlton, pay very close attention to understanding and continuously improving this interface, and it results in customer loyalty. Companies can analyze the customer interaction process, measure results in terms of customer satisfaction, and learn how to improve that process.

> For more information, see: Valarie A. Zeithaml, A. Parasuraman, and Leonard L. Berry, *Delivering Quality Service* (New York: The Free Press, 1990); Thomas H. Berry, *Managing the Total Quality Transformation* (New York: McGraw-Hill, 1991).

Key Performance Indicator (KPI) These are certain aspects of performance that quantitatively show that a company is improving the efficiency of its processes and the quality of its outputs. Implementing quality management includes creating quantitative measures of KPIs that will provide data to track organization performance.

Examples of KPIs include accidents on the job, costs for specific processes, time to complete each process step, amount of work-in-process, measures of customer satisfaction, on-time delivery percent, and so on. While measures of all processes are important for improvement, the creation of KPIs gives everyone in the organization an immediate picture of how they are doing. In creating KPIs, it is important that they be shared with all employees.

> For more information, see: James M. Carman, "Continuous Quality Improvement as a Survival Strategy: The Southern Pacific Experience," *California Management Review*, Spring 1993, pp. 118-132.

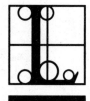

Lagging Measure This is a measure or result issued at the end of a process, event, or time period.

This is the traditional form of measurement and reporting and is still the most widely used. Reporting sales at the start of the month for the previous month is an example. So is the reporting of how many widgets were made last week. Such measures are important because they help us know how a company is doing. However, they also focus on ends not means. They emphasize making numbers not making improvements in the processes that deliver the results. Implementing TQM suggests that we must supplement these types of lagging measures with in-process measures that give immediate feedback on performance, process variation, special cause problems, and how to make improvements that will improve quality, lower costs, satisfy customers, and facilitate growth in sales and profits.

For further information, see: Steven M. Hronec, *Vital Signs: Using Quality, Time, and Cost Performance Measurement to Chart Your Company's Future* (New York: Amacom, 1993); Jacek Koronacki and James R. Thompson, *Statistical Process Control for Quality Improvement* (New York: Chapman & Hall, 1993).

Leaders/Leadership Leaders shape and influence the principles, values, and methods by which an organization achieves its purposes and goals. Leaders, as opposed to bosses, create an environment that brings out the best in employees. Leadership includes being able to tap into the intrinsic motivation in people and help them identify their welfare with that of the organization. Leaders are fair and consistent in the practice of espoused principles and values.

TQM recognizes the vital role organizational leaders play in shaping values, such as open communication, focus on processes, teamwork, continuous improvement, and customer delight that will result in happy customers, a stimulating work environment, and a growing, profitable company. Every major study of cultural transformation and the migration of a company or agency to a quality-driven or customer-focused culture recognizes the importance of leadership. The main reason TQM fails in companies is because company senior managers do not practice the leadership necessary to instill the values and principles on which it is based. These managers (not leaders) view TQM as a quick fix and do not recognize the profound shift it requires away from traditional hierarchical management values and practices. In those organizations where TQM is producing results, conversely, we see a deep and continuing commitment to quality values and practices by leaders that transcends individuals and envelops the entire organization. It brings out a kind of company-wide leadership at all levels of the enterprise, which we now identify as empowerment. Both personal leadership and empowerment across the organization are necessary for implementing quality principles and practices.

> For further information, see: Peter M. Senge, *The Fifth Discipline* (New York: Doubleday Currency, 1990); Richard S. Johnson, *TQM: Leadership for the Quality Transformation* (Milwaukee, WI: ASQC Quality Press, 1993); Joan E. Gebhardt and Patrick Townsend, *Quality in Action: 93 Lessons in Leadership, Participation, and Measurement* (New York: John Wiley & Sons, 1992); William Lareau, *American Samurai* (New York: Warner Books, 1991); Jack Stack, *The Great Game of Business* (New York: Doubleday Currency, 1992).

Leading Measures These are measures that take place at different steps while a process is underway.

> In a learning organization, leaders are designers, stewards, and teachers. They are responsible for *building organizations* where people continually expand their capabilities, clarify vision, and improve shared mental models.
>
> *Peter M. Senge*

At a minimum, establish a measurement system supporting process performance for those critical subprocesses or activities. The ultimate goal is is to establish measurement and feedback systems for every activity within a process.

H. James Harrington

To understand fully what is happening in the course of a process, measures are established for each important step. Those used early or in the initial steps are called leading measures. These give a process owner an early indication of how well a process is performing. These measures also serve as an early warning of a problem that may increase the variation and nonconformance of outputs further down the line. They can then deal with such problems as they occur. For example, an indicator of a poorly designed part, before it went into production, could suggest to a manufacturing engineer that, once produced, the part might cause a recall of the product. An indicator that is used to monitor the quality design of a pollution control device on a car can save an auto manufacturer the cost and embarrassment of having to recall hundreds of thousands of vehicles for a faulty emission control system two years later!

> For more information, see: Eugene H. Melan, *Process Management: Methods for Improving Products and Services* (New York: McGraw-Hill, 1993); H. James Harrington, *Business Process Improvement* (New York: McGraw-Hill, 1991).

Lead Team This is the team that guides or supervises other process teams. Frequently the term is used to describe the senior managers of an organization and their role in guiding other teams.

Another common phrase is "champions of change" or, simply, "management team." Regardless of the term used, the concept underlying a lead team is that management acts like a team, not simply as heads of individual functional organizations representing their own interests. This calls for understanding the importance of cross-functional cooperation as well as looking at management as a process that includes its own steps, methods, and opportunities for learning and improvement.

> For more information, see: Alexander Hiam, *Closing the Quality Gap: Lessons from America's Leading Companies* (Englewood Cliffs, NJ: Prentice-Hall, 1992); V. Daniel Hunt, *Managing for*

Quality (Burr Ridge, IL: Irwin Professional Publishing, 1993); David A. Garvin, *Managing Quality: The Strategic and Competitive Edge* (New York: The Free Press, 1988).

Lessons Learned In quality practice this phrase refers to identifying facts of the behavior of processes that can be applied to the future management of those processes.

Joseph M. Juran popularized the phrase. He suggests that three tools result from actively using this idea that take advantage of learning from experience to help make future operations work better. These tools are:

- *Data bank,* which is a compilation of facts organized for easy retrieval. Computers simplify the use of these.
- *Checklists,* which serve as aids to memory to help remember what to do, what needs to be done, and sometimes what to avoid.
- *Countdown,* which is a list of items to do in a particular sequence to properly execute a process.

In his book, *Juran on Leadership for Quality,* he suggested the following benefits from using such tools of lessons learned:

- They make available to any user the collective experience and memories of many individuals, organized in ways that permit ready retrieval.
- They are of a repetitive nature: They can be used over and over again for an indefinite number of planning cycles.
- They are impersonal. They avoid the problems created when one person gives orders to others.

For more information, see: Joseph M. Juran, *Juran on Leadership for Quality: An Executive Handbook* (New York: The Free Press, 1989); Joseph M. Juran, *Juran on Planning for Quality* (New York: The Free Press, 1988).

Level of Seriousness This refers to the degree or level of possible negative effects that a defect or other lack of conformance to

> Collectively, human use of the concept of lessons learned has been decisive in human dominance over all other animal species. So awesome a result suggests that the degree of use of lessons learned can also play a significant role in competition in the marketplace.
>
> *Joseph M. Juran*

specifications of components, products, or processes may cause for the company or the user.

Any classification scheme that categorizes or ranks degree of defects or problems would incorporate some procedure that documents level of seriousness. The normal approach is to rank a problem from 0 to 100 percent or 1 to 10, based on some predetermined characterization of each value. This ranking of problems is particularly useful in defining which issues to work on first, following, in effect, the Pareto principle that the majority of problems are caused by a critical few serious defects. (See also *Defect*.)

> For more information, see: Armand V. Feigenbaum, *Total Quality Control*, Third Edition, Revised (New York: McGraw-Hill, 1991).

Leverage In management, this idea suggests understanding an organization and its processes so that actions taken will yield the greatest possible good for the enterprise and for customers.

The concept of leverage has special value when viewing an organization as a system. For example, when you identify a bottleneck in a process, taking action to open it up can have a much more positive effect throughout the entire process than any other action you might take. And it can be relatively simple to make that change. In any process there is usually a point where you can intervene and have the greatest possible effect for the amount of time, money, and effort expended.

The Pareto principle, or the 80/20 rule, for example, helps identify leverage points. Other TQM tools for gathering data and analyzing processes also help managers decide what actions will give the greatest return. A problem with traditional management is that it does not focus on leverage points, and many actions taken are just as likely to make a problem worse as to improve it.

> For more information, see: Brian L. Joiner, *Fourth Generation Management* (New York: McGraw-Hill, 1994); Peter M. Senge, *The Fifth Discipline* (New York: Doubleday Currency, 1990).

Time and again I've encountered organizations, divisions, departments, and even individuals who deceive themselves into thinking they can manage 20 priorities. But by having 20 "priorities," we end up having none. Organizations that can focus on the one, two, or perhaps three most important leverage points have much more success.

Brian L. Joiner

Life Cycle Cost (LCC) This refers to the total cost of a system or product over its useful life. These costs include acquisition, ownership, and final disposal.

For example, suppose a computer costs $400,000 and will be used for five years. Maintenance costs and the expense of the staff using the machine might cost an additional $400,000 each year. The life cycle cost of this machine would be $2.4 million. If using ABC accounting techniques, the various activities done with the machine can be identified as expenses over the life of the machine's utilization. In other words, life cycle costs are the total amount of expenses incurred by the purchaser for as long as he or she uses the machine or any piece of equipment or other resource. In the world of quality, total costs are studied with an eye toward improving performance of a product to lower total costs to consumers while gaining competitive advantage over a rival.

For more information, see: Joseph M. Juran and Frank M. Gryna, *Quality Planning and Analysis*, Third Edition (New York: McGraw-Hill, 1993); Armand V. Feigenbaum, *Total Quality Control*, Third Edition, Revised (New York: McGraw-Hill, 1991); Peter B.B. Turney, *Common Cents: The ABC Performance Breakthrough* (Portland, OR: Cost Technology, 1991).

Line Chart This kind of chart displays and compares quantifiable data, usually displaying changes over time by showing a line running from point to point. It is also known as a run chart. Line or run charts are one of the basic tools of TQM.

Once data is plotted on a line chart, you can then calculate and enter control limits to learn how a system is operating and whether measurements indicate if an output is in or out of statistical control. Figure 50 (p. 206) shows a line chart for inches of snow over a period of four weeks.

For more information, see: Michael Brassard and Diane Ritter, *The Memory Jogger II* (Methuen, MA: GOAL/QPC, 1994); N. Logothetis, *Managing For Total Quality: From Deming to Taguchi and SPC* (Englewood Cliffs, NJ: Prentice Hall, 1992); Hy Pitt, *SPC for the Rest of Us* (Reading, MA: Addison-Wesley, 1994).

FIGURE 50

A line chart showing inches of snowfall during five weeks of winter.

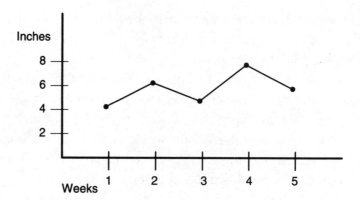

Linkage Analysis This is a process tool or methodology for identifying internal and external linkages to determine activities, processes, or organizations that are highly dependent on one another.

In manufacturing, these could involve identifying dependencies on suppliers of components. In a community, it could identify charitable agencies on which the city depends to do work that it would otherwise have to do. Linkage analysis is a useful perspective in thinking through such issues as:

- Who should a company partner with?
- Who are the company's suppliers?
- Who are the company's stakeholders?
- Who can and should help the company?

For more information, see: Kenneth Primozic, Edward Primozic, and Joe Leben, *Strategic Choices* (New York: McGraw-Hill, 1991); Michael E. Porter, *Competitive Advantage: Creating and Sustaining Superior Performance* (New York: The Free Press, 1985).

Linkages These are relationships between interdependent activities in a value chain or between interdependent organizations. Linkages are often said to exist if the performance of one process or organization has a direct or indirect influence on the performance of other processes or organizations.

One application of this concept is in the development of the critical path for a project or process. Identifying the linkages between operations shows which ones must be completed according to specifications before others can be completed. Failure to understand the linkages among processes is an important cause of waste and inefficiency in an organization. This raises costs and lowers performance.

> For more information, see: Michael E. Porter, *Competitive Advantage: Creating and Sustaining Superior Performance* (New York: The Free Press, 1985); Kenneth Primozic, Edward Primozic, and Joe Leben, *Strategic Choices* (New York: McGraw-Hill, 1991).

Listening This is the act of receiving, understanding, and reacting to information, such as customer interests and concerns. In customer feedback processes, the term is often used in such contexts as the establishment of "listening posts."

In process work and quality-driven management, listening consists of two actions. One comprises all processes employed to ensure that everyone in the organization is hearing the "voice of the customer" and applying what they hear to operating processes that result in outputs that meet customer requirements. A second collection of listening activities involves paying attention to the thoughts and ideas of employees. These range from morale surveys to suggestion processes—all critical to taking advantage of the knowledge and dignity of all employees.

> For more information, see: Bob E. Hayes, *Measuring Customer Satisfaction* (Milwaukee, WI: ASQC Quality Press, 1992); Kathleen D. Ryan and Daniel K. Oestreich, *Driving Fear Out of the Workplace* (San Francisco: Jossey-Bass, 1991); James N. Salter II and J. Stephen Sarazan, *Customer Satisfaction Management Process* (New York: American Management Association, 1993); Peter M. Senge et al., *The Fifth Discipline Fieldbook* (New York: Doubleday Currency, 1994); Tom Hinton and Wini Schaeffer, *Customer-Focused Quality* (New York: Prentice Hall, 1994).

Listening Posts These are the points either in a process or physically in a building where data is accumulated on the performance and quality of processes, products, or services.

Examples of listening posts can vary widely. For instance, the last inspection point on a production line is a listening post. So

> Although activities are the building blocks of competitive advantage, the value chain is not a collection of independent activities but a system of interdependent activities. Various tasks are related by linkages within the value chain. Competitive advantage derives from *linkages among activities* just as it does from the individual activities themselves.
>
> *Michael E. Porter*

> To listen fully means to pay close attention to what is being said beneath the words. You listen not only to the "music," but to the very essence of the person speaking. You listen not only for what someone knows, but for who he or she is.
>
> *William Isaacs, from The Fifth Discipline Fieldbook*

too is that point in a computer program where information is gathered on the time it takes for a transaction to occur. Many companies are establishing listening posts at several points in their processes to serve as early warning signals of quality problems so that they can be resolved before a product or service deteriorates. Listening posts are also used to assess the overall performance of processes. Traditionally, listening posts are the points where inspection takes place. However, by emphasizing the idea of listening rather than inspection, it focuses everyone on listening to the "voice of the process" and working on preventing problems rather than inspecting the work of individuals.

> For more information, see: Tom Hinton and Wini Schaeffer, *Customer-Focused Quality* (New York: Prentice Hall, 1994).

Little c Customers   This phrase reminds all employees in a company that they are all customers (and suppliers) to others in the company.

The concept behind this phrase is that all employees should produce quality workmanship that meets the requirements of fellow employees who depend on this workmanship. Without doing this for the next customer down the line in a process, it becomes difficult for that customer (employee) to perform his or her work at an appropriate level of quality. For example, if a sales representative brings in an order without all the particulars required by the company, the order entry clerk in the office cannot do his or her job completely, thereby performing below minimum requirements. Conversely, if the sales representative was attentive to the needs of the order entry clerk (and thus to the process as a whole), then the order entry clerk would have all the information required to complete the order entry process. The benefits are obvious: The work gets done quicker, more accurately, and, using our example, the order is fulfilled sooner.

> For more information, see: James W. Cortada and John A. Woods, *The Quality Yearbook* (New York: McGraw-Hill, 1994, 1995); Richard C. Whiteley, *The Customer-Driven Company* (Reading: Addison-Wesley, 1991).

Little Q This is a narrow scope of quality, covering only the idea of a product conforming in some defined way to customer expectations or requirements. In other words, it focuses on the final output of a company and not on the processes by which that output gets delivered. (See also *Big Q*)

For more information, see: H.G. Menon, *TQM in New Product Manufacturing* (New York: McGraw-Hill, 1992); Joseph M. Juran, *Juran on Leadership for Quality* (New York: The Free Press, 1989).

Loss Function This is a way of analyzing and understanding how variations from customer-designated quality characteristics represent costs to the organization. It is a formal way of representing the "cost of poor quality."

It is a key component of the Taguchi approach to the improvement of any system or process and is often known as the "Taguchi loss function." The basic assumption is that any variation from a target is considered a loss. Losses generate additional costs, and thus the further from a target that the performance of a process or system is, the greater the expense of that process. It is often used in manufacturing settings.

In the past, this concept has characterized a difference between the American and Japanese approaches to parts manufacturing. The American approach has been to set a range of values for a specification in determining the acceptability of any part. This is sometimes known as the "goal post" approach, taken from football and meaning that any value that falls within the goal posts is acceptable and any that fall outside represent a loss. The Japanese approach was to set an exact value for any specification. Any variation from that value would represent a loss because it would increase the possibility of failure in the final product due to imprecise fitting among parts.

For example, if the measured diameter of a shaft that is to fit within a machined hole in an engine is acceptable within a few hundredths of an centimeter, there is always the possibility of tightness or looseness that will adversely affect the performance of the engine. On the other hand, a shaft that meets the

> The loss function provides a way to quantify the potential savings that can be achieved by reducing variation around a target value. This is in contrast to the notion that everything within specs is equally good and everything outside of specs in equally bad.
>
> *Ronald M. Fortuna, from Total Quality: An Executive's Guide for the 1990s*

precise specification is less likely to fail or cause problems. Understanding this, taking the Taguchi approach drives everyone in the company to look for possible causes of variation and to adapt measurements and methods for continuously reducing the amount of this variation. This helps ensure that the final outputs will function properly and minimize the possibility of failure.

We can use this same idea to understand that any variation from customer specifications represents a loss as well. It helps move managers from an attitude of generally giving customers what they expect to one of continuously improving a company's ability to precisely meet and even exceed the customers' expectations. In other words, it suggests always delivering products that minimize loss to customers and, thus, to the company as well. Figures 51 and 52 show the traditional and Taguchi methods of illustrating a loss function.

For more information, see: Phillip J. Ross, *Taguchi Techniques for Quality Engineering* (New York: McGraw-Hill, 1988); Brian L.

FIGURE 51

Traditional approach to quality. Anything that measures within goal posts is acceptable.

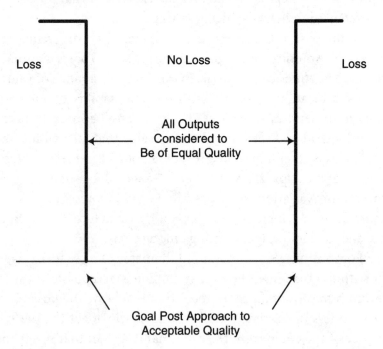

Loss　　　　　No Loss　　　　　Loss

All Outputs
Considered to
Be of Equal Quality

Goal Post Approach to
Acceptable Quality

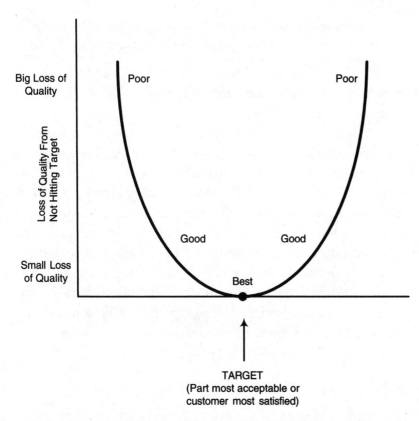

Big Loss of
Quality

Loss of Quality From
Not Hitting Target

Poor   Poor

Good   Good

Small Loss
of Quality   Best

TARGET
(Part most acceptable or
customer most satisfied)

FIGURE 52

Taguchi loss function.
The farther from the tar-
get as specified by the
customer or the specifi-
cation, the more loss
and the compromise of
quality and value.

Joiner, *Fourth Generation Management* (New York: McGraw-Hill, 1994); Ernst & Young, *Total Quality: An Executive's Guide for the 1990s* (Burr Ridge, IL: Irwin Professional Publishing, 1990).

Lot This is a specific quantity of products or outputs from a process. The items in a lot are considered uniform for the purpose of judging their overall quality by sampling only a portion of them.

The idea of item uniformity is crucial to understanding the lot concept. If all pieces in a lot are uniform, then a random sampling of units should give an accurate reading of the quality of all units.

For more information, see: Joseph M. Juran and Frank M. Gryna, *Quality Planning and Analysis*, Third Edition (New York: McGraw-Hill, 1993); Armand V. Feigenbaum, *Total Quality Control*, Third Edition, Revised, (New York: McGraw-Hill, 1991); George

L. Miller and LaRue L. Krumm, *The Whats, Whys & Hows of Quality Improvement* (Milwaukee, WI: ASQC Quality Press, 1992).

Lot Plot Plan This is a sampling plan that employs histograms to help make acceptance/rejection decisions on entire lots of components or products.

The normal practice is to randomly pick samples of 50 products or components of the same type from a lot. These samples are measured for degrees of variability and then divided into as many as 10 different cells. Through the use of statistics, one can calculate degree of variability in the entire lot based on this sample. Those lots that fall within specifications are accepted, and those that do not are rejected. This kind of data also guides a process or product owner in defining defects in operations and quality, leading to improvements in overall performance. This technique can also lead to decisions about which suppliers are the best. Figure 53 shows an example of a lot plot histogram.

FIGURE 53

An example of a lot plot histogram showing that all variation in samples from a lot falls within specification limits.

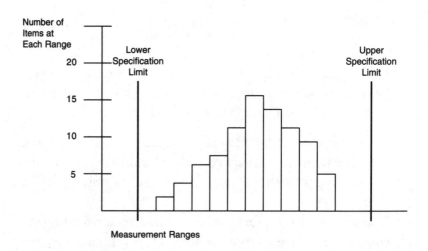

For more information, see: Eugene L. Grant and Richard S. Leavenworth, *Statistical Quality Control*, Sixth Edition (New York: McGraw-Hill, 1988).

Lot Tolerance Percent Defective (LTPD) This is a way of quantitatively defining the relationship between the probability of acceptance by a customer of a product lot compared to the possible percent of defectives within that lot.

This relationship shows that the higher the percent of defectives, the lower the probability of accepting the lot. LTPD is important because in acceptance sampling plans, not every item in a lot is inspected. This means there is always the possibility that the sampling plan will not properly represent the actual percent of defectives in a lot. Understanding this, the goal is to develop a sampling plan to minimize this possibility. However, no such plan except 100 percent sampling can do this. Thus, statisticians have come up with a standard quantifiable means of representing this relationship. Acceptable quality level might be a 95 percent probability that lots of products will contain not more than 5 percent defectives. In other words, 95 percent of the time, given any particular sampling plan, customers will not accept a lot with more than 5 percent defectives. This method also helps assure that the sampling plan will not allow more than a 10 percent probability of a lot being accepted with more than 8 percent defectives (or whatever the outside limit is on defective percent). These percentages of defects may be different depending on acceptable quality level. See *Operating Characteristic Curve* for more of this concept.

> For more information, see: George L. Miller and LaRue L. Krumm, *The Whats, Whys & Hows of Quality Improvement* (Milwaukee, WI: ASQC Quality Press, 1992); Joseph M. Juran and Frank M. Gryna, *Quality Planning and Analysis*, Third Edition (New York: McGraw-Hill, 1993).

Lower Control Limit (LCL) This is the line on a control chart indicating the lower limit within which a process is in statistical control. The LCL is positioned on the chart three standard deviations below the average of the measurements of process outputs through time.

Any measurement of process performance that falls between the lower and the upper control limits is due to common cause variation that is inherent in the system. Any attempt to treat a

A process may be in statistical control yet turn out 10 percent defective—10 out of 100 items outside specifications. In fact, a process could be in statistical control yet turn out 100 percent defective. There is no logical connection between control limits and specifications. Control limits, once we have achieved a fair state of statistical control, tell us what the process is and what it will do tomorrow. The control chart is the process talking to us.

W. Edwards Deming

FIGURE 54

A typical control chart with the lower control limit labeled. Process is in statistical control.

single data point that falls within control limits as if it were a special cause (a common management problem) will not help the system and only introduces new sources of variation. If performance is above the upper control limit or below the lower control limit, this indicates that the variation noted is due to a special cause, which we can usually identify and quickly do something about. Figure 54 shows a control chart with the lower control limit.

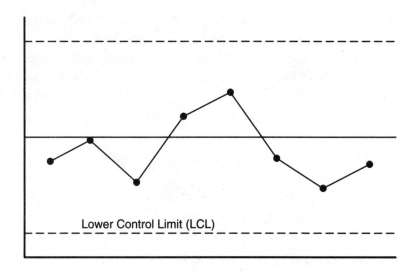

Lower Control Limit (LCL)

For more information, see: W. Edwards Deming, *The New Economics* (Cambridge, MA: MIT Center for Advanced Engineering Study, 1993); William W. Scherkenbach, *Deming's Road to Continual Improvement* (Knoxville, TN: SPC Press, 1991); Brian L. Joiner, *Fourth Generation Management* (New York: McGraw-Hill, 1994).

LTPD This is the acronym for lot tolerance percent defective.

Macro Activity In ABC accounting, this refers to an aggregation of related activities, such as manufacturing, sales, or distribution. By aggregating related activities in this manner, ABC can still attribute costs to the proper activity without resorting to more detail than is necessary.

> For more information, see: Peter B.B. Turney, *Common Cents: The ABC Performance Breakthrough* (Portland, OR: Cost Technology, 1991).

Macro-Process This is a major process of an enterprise that is made up of numerous other processes (sometimes called subprocesses) forming an overall system of activities within an enterprise. One way of understanding this is as a hierarchy, with the macro-process at the top of the hierarchy, followed by subprocesses, activities, and tasks.

Macro-processes often involve multiple departments and functions. They also consist of many processes that are managed independently but which all share a dependency on each other. For example, the "manufacturing process" is a macro-process. Within that broad activity are a variety of more narrowly focused yet interdependent processes: design, engineering, purchasing, and production. Each of these involves various activities and, at the micro level, tasks. Figure 55 illustrates this concept.

FIGURE 55

Macro-process with subprocesses, activities, and tasks.

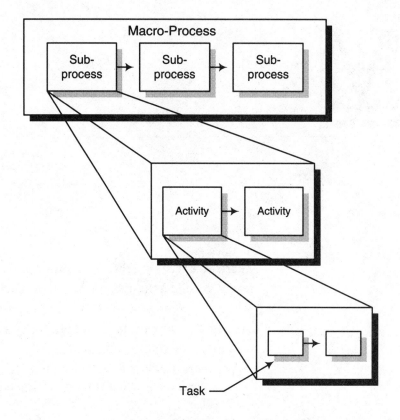

For more information, see: William Lareau, *American Samurai* (New York: Warner Books, 1991); H. James Harrington, *Business Process Improvement* (New York: McGraw-Hill, 1991).

Maintainability This is a manufacturing term that refers to the characteristics of the design or installation of a piece of equipment that makes it possible to later repair it easily and efficiently. The degree to which a product or device can be easily maintained is at the heart of this concept.

Often replacement products for an earlier generation of similar goods are designed to improve maintainability and to lower costs of support. An example would be newer generations of aircraft. Maintainability is often a consideration in determining when to replace one piece of equipment with another. This is because in any assessment of the total costs of a device, you must take into account maintenance costs. For example, a computer

Reliability and maintainability are two of the primary drives of availability of the product for intended use when it is needed. R&M, along with logistics considerations, are major contributors to customer satisfaction.

James H. Saylor

system that costs $1 million and requires another $1 million in maintenance expenses over five years is more expensive than a $1.5 million system that calls for $400,000 in maintenance. More complicated software packages and programming languages often are subject to discussions about maintainability. So, too, are machines that require extensive set up and tear down time, or extensive maintenance (for example, paper-making equipment). From the point of view of a product's manufacturer or a process owner, designing in maintainability is an important consideration.

For more information, see: Joseph M. Juran and Frank M. Gryna, *Quality Planning and Analysis*, Third Edition (New York: McGraw-Hill, 1993); Armand V. Feigenbaum, *Total Quality Control*, Third Edition, Revised (New York: McGraw-Hill, 1991); Eugene R. Carrubba and Ronald D. Gordon, *Product Assurance Principles* (New York: McGraw-Hill, 1988); James H. Saylor, *TQM Field Manual* (New York: McGraw-Hill, 1992).

Maintenance This is always defined as the repair or preservation of a device or process at a predetermined level of quality. In general, it refers to those activities required to keep a piece of equipment running properly.

In quality management, maintenance is subject to extensive statistical process control to determine what constitutes acceptable performance by, for example, a piece of equipment. The activities performed to maintain that machine are part of a process, which also can be documented, maintained, and improved.

For more information, see: Joseph M. Juran and Frank M. Gryna, *Quality Planning and Analysis*, Third Edition (New York: McGraw-Hill, 1993).

Making the Rounds In quality management, this is the act of visiting customers to obtain their thoughts on how to improve products or services. We also use the phrase to describe the act of visiting other departments or teams in a company that will be affected by those actions currently being planned.

Making the rounds is particularly important if you are modifying a process that involves or affects other parts of your company or organization. It is possible, for example, to improve a process from your point of view and, by not making the rounds, negatively affect the productivity or quality of the work of others in the organization.

For more information, see: William Lareau, *American Samurai* (New York: Warner Books, 1991).

Malcolm Baldrige National Quality Award This is the award established by the U.S. government in 1987 to recognize quality achievements by American companies. It is awarded by the U.S. Department of Commerce and is today the most prestigious quality award in the United States.

The basic concept behind the award is that there are seven sets of quality criteria that, when properly implemented, lead an organization to perform in a superior manner—reducing waste and inefficiencies, creating a healthy workplace, and successfully serving customers. There are 1,000 possible points distributed among the seven criteria groups. A panel of judges appraises firms applying for the award and awards points according to how well these firms meet each set of criteria.

The criteria along with the points as of 1995 fall into the following categories:

1. Leadership, 90 points.
2. Information and Analysis, 75 points.
3. Strategic Planning, 55 points.
4. Human Resource Development and Management, 140 points.
5. Process Management, 140 points.
6. Business Results, 250 points.
7. Customer Focus and Satisfaction, 250 points.

The Baldrige Criteria has become the basis for many U.S. state quality awards and many corporate quality strategies. The criteria serve as a valuable auditing device for any company seeking to judge where it now stands in implementing TQM cultural values and practices. Figure 56 shows the interdependency of the seven Baldrige Award Criteria.

> Of all the Baldrige Award program elements, the most valuable is the Baldrige criteria, which organizations use to assess their quality systems. The fact that such an all-encompassing definition never existed before is one reason American managers have had so much trouble getting their arms around quality.
>
> *Stephen George*

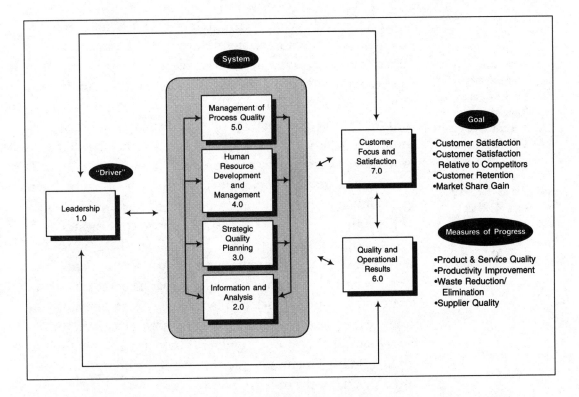

For more information, see: Mark Graham Brown, *Baldrige Award Winning Quality* (White Plains, NY: Quality Resources and Milwaukee, WI: ASQC Quality Press, Annual); Stephen George, *The Baldrige Quality System* (New York: John Wiley & Sons, 1992); Christopher W.L. Hart and Christopher E. Bogan, *The Baldrige: What It Is, How It's Won, How to Use it To Improve Quality in Your Company* (New York: McGraw-Hill, 1992); Donald C. Fisher, *Measuring Up to the Baldrige* (New York: Amacom, 1994); James W. Cortada and John A. Woods, *The Quality Yearbook* (New York: McGraw-Hill, Annual). To get the current year's criteria for the Malcolm Baldrige National Quality Award, write to: National Institute of Standards and Technology, Route 270 and Quince Orchard Road, Administrative Building, Room A537, Gaithersburg, MD 20899-0001, telephone: (301) 975-2036, fax: (301) 948-3716.

FIGURE 56

The Baldrige Award framework, showing the interrelationships of the seven criteria, all of which feed into the goal of customer satisfaction, which feeds back into the system.

Management This is the collective leadership of an enterprise. Management sets the goals, marshals the resources, and provides the direction for organizational action.

Every employee has elements of management in his or her job. We must have all employees, to the level that's appropriate to their duties, teaching and coaching associates, seeking innovations in their work, planning for better processes, and seeking to understand and reduce variability.

William Lareau

From a quality perspective, management is no longer the private preserve of a class of employees called managers but is shared among all employees. Managers and management in these companies take on the role of leaders and coaches responsible for training, facilitating change, removing roadblocks to employee performance, and providing resources. For their part, employees take on more responsibility for making decisions about how processes will perform, their improvement, the setting of goals, and the monitoring of process performance. Managers and employees alike often operate as members of teams, with training in teamwork, statistical process control, and overall skill improvement. Successful management behavior in this new context places a high premium on cooperation and communication and everyone working together in an informed way. The overall responsibility of management is to continuously improve the processes by which the organization creates and delivers high-quality products and services to its customers. In this context, management sets the direction, including the vision and mission, for the organization.

In many well-run companies, management still consists of traditional activities such as planning, organizing, and controlling, but with much more delegation of work to process teams, often consisting of members from several functional areas. Figure 57 illustrates a modern view of management responsibilities. These include bringing together resources, developing and measuring processes, and facilitating the decisions that maximize and continuously improve customer satisfaction. In this view, management ensures that the company hears the voices of its customers and all its stakeholders, identifies and works out problems and opportunities, and makes key strategic decisions. Process teams perform the key tasks of the company, drawing on the organization for resources (for example, people, culture, and information technology). At each point in a process cycle, managers gather and study measurements and use these to make decisions and guide process and quality improvement. To a growing extent,

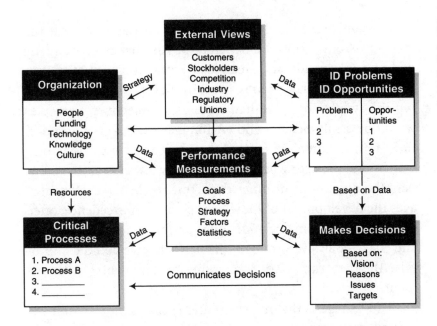

FIGURE 57

A schematic showing the responsibilities and relationships that a manager must be concerned with.

companies are using ISO 9000 audits and the Baldrige Criteria assessments to document how well management is doing.

> For more information, see: David A. Garvin, *Managing Quality* (New York: The Free Press, 1988); Peter R. Scholtes, *The Team Handbook* (Madison, WI: Joiner Associates, 1988); John P. Kotter and James L. Heskett, *Corporate Culture and Performance* (New York: The Free Press, 1992); James. C. Collins and Jerry I. Porras, *Built to Last: Successful Habits of Visionary Companies* (New York: HarperBusiness, 1994); Brian L. Joiner, *Fourth Generation Management* (New York: McGraw-Hill, 1994); William Lareau, *American Samurai* (New York: Warner Books, 1991); James W. Cortada, *TQM for Information System Management* (New York: McGraw-Hill, 1995).

Management by Objectives (MBO) This is a standard management practice in which a manager and employee set quantifiable objectives to meet within the next time period. Then the manager judges the success of the employee based on whether he or she has achieved these objectives.

MBO is generally looked at from a TQM perspective as flawed. This is because it focuses on individual performance rather than processes. It does not take into account that the success of any employee depends on the behavior of many other people inside and outside the organization. MBO encourages individuals to look good to their boss by manipulating the system to make their numerical objectives. Examples would be a salesperson who encourages customers to send in their returns after the end of the company's fiscal year or a warehouse manager who keeps inventory low even if that hampers the operations of other departments. MBO also encourages the development of individual pockets of excellence, often achieved by optimizing individual performance to look good and make goals even though this performance could compromise the performance of some other person, team, or department.

TQM suggests that managers and everyone in the company sign on to objectives that focus on process improvement and enhanced customer satisfaction. Then top management makes sure everyone has the training, tools, and authority so they can work together to achieve such objectives. Further, we see the abandonment of the appraisal of individuals by whether they have met objectives or not. The emphasis is on team performance and successful interaction among employees to execute and improve processes. The company also seeks not the performance optimization of individuals or departments, but optimization of the system as a whole. All of this is consistent with Point 11b of Deming's 14 Points: "Eliminate management by objective. Eliminate management by numbers, numerical goals. Substitute leadership."

> For more information, see: Brian L. Joiner, *Fourth Generation Management* (New York: McGraw-Hill, 1994); W. Edwards Deming, *Out of the Crisis* (Cambridge, MA: MIT Center for Advanced Engineering Study, 1986); Peter F. Drucker, *Management: Tasks, Responsibilities, Practices* (New York: Harper & Row, 1973); Aubrey C. Daniels, *Bringing Out the Best in People* (New York: McGraw-Hill, 1994); William Lareau, *American Samurai* (New York: Warner Books, 1991).

Manufacturing Resource Planning (MRP) This refers to a collection of planning processes that revolve around manufacturing. These processes include scheduling, inventory management, work orders, work order tracking, and labor reporting. Historically, this has been an important method for planning and controlling manufacturing resources and operations.

Though MRP is a well-established, widely used manufacturing planning tool, there has always been controversy concerning its effectiveness. In recent years, for example, just-in-time (JIT) approaches have challenged MRP by emphasizing more synchronization of all activities, unlike MRP, which stresses plans that mainly focus on details of individual shop-floor activities. Increasingly, MRP-based systems have incorporated some JIT functions such as rate-based planning. (See also *Just-in-Time Manufacturing*.)

> For more information, see: Richard J. Schonberger, *World Class Manufacturing* (New York: The Free Press, 1987); H.G. Menon, *TQM in New Product Manufacturing* (New York: McGraw-Hill, 1992).

Margin In quality management, this refers to the difference between total value of an output and the combined costs of executing the processes and activities that result in that output.

Market-Driven Quality (MDQ) This is IBM's term for Total Quality Management (TQM). It incorporates the idea that the company will use the market's needs to help define its strategy and how it will allocate its resources. It is the acknowledgment that focusing on customer satisfaction provides the foundation for managerial decisions and actions as the company seeks to deliver defect-free goods and services. Further, it suggests that in every aspect of the firm's operations it must strive to improve its performance in doing this.

The term is increasingly being used by other companies in similar ways. Customer-driven inspiration for a company's products, services, and measures of success are frequently seen as market driven.

For more information, see: James W. Cortada, *TQM for Sales and Marketing Management* (New York: McGraw-Hill, 1993); Roy A. Bauer, Emilio Collar, and Victor Tang, *The Silverlake Project* (New York: Oxford University Press, 1992).

Market-In This is a term describing an organizational consciousness and culture in which the market's needs and expectations define and give meaning to the decisions and actions of everyone in the company. This is also known as "customer-in."

It suggests the development of a culture where all employees are aware of their role in adding value to the company's final outputs. The idea of market-in focuses managers and all employees on understanding and defining their work in terms of customer satisfaction. Most importantly, it is a cultural value in which the market—customers—are primary in understanding what the company does and what motivates the behavior of everyone in the company.

> The "customer-in" organization bonds supplier to customer in a multitude of ways. Since every customer is also a supplier to another customer, a chain of customers, each well connected to the next, is the goal.
>
> *Richard J. Schonberger*

For more information, see: Richard J. Schonberger, *Building a Chain of Customers* (New York: The Free Press, 1990).

Mass Customization A relatively new concept in manufacturing, this is the notion that one can create variety and customization through flexibility and quick response in manufacturing processes. In other words, it is a body of product design and manufacturing processes that allow one to, in effect, make one-of-a-kind products while using the techniques and enjoying the economic benefits of mass production.

> Practitioners of Mass Customization share the goal of developing, producing, marketing, and delivering affordable goods and services with enough variety and customization *that nearly everyone finds exactly what they want.*
>
> *B. Joseph Pine II*

The approach allows a company to produce a variety of models and colors of automobiles on one assembly line, for example, or potentially publish this book in a variety of editions tailored to different customers and their needs. Mass customization prizes flexibility and quick responsiveness in design and production. It delivers exactly what customers want. It assumes that all demand is fragmented and that markets are niches. It incorporates the understanding that success calls for developing the capacity to operate short production runs at costs comparable to mass production with no compromise in quality. It further suggests that products are likely to have short life cycles, and the company must continuously upgrade its ability to serve individual customers.

> For more information, see: B. Joseph Pine II, *Mass Customization: The New Frontier in Business Competition* (Boston: Harvard Business School Press, 1993).

Materials Review Board (MRB) This is a quality control committee or team, usually employed in manufacturing or other materials processing installations. This board has the purpose of determining how to deal with materials that do not conform to predetermined standards of quality or to make sure that components used in manufacturing conform to "fitness for use" specifications.

Members of an MRB typically include process-control engineers, designers, purchasers of materials, and quality inspectors, who have the goal of assuring that their organization uses materials that meet specific standards of quality. In practice, they make a fitness-for-use decision about the viability of a component or material. This decision-making process is also part of the larger effort of establishing and maintaining quality standards from suppliers.

> For more information, see: Joseph M. Juran and Frank M. Gryna, *Quality Planning and Analysis*, Third Edition (New York: McGraw-Hill, 1993).

Matrix Organization This is a form of organization of a company or agency in which a team structure is typically superimposed on a functional organization. It may also be an organization in which functional departments work closely with other departments on joint projects and processes and are held accountable for results of their joint work. The matrix organization is a critical requirement for the successful implementation of the TQM concept of cross-functional teams.

This kind of organization has become increasingly popular with corporations. A general manager in one division would, in this kind of structure, be responsible for the development and health of a process that cuts across the functional departments of other general managers. Work done by one organization would become input to the activities of other departments. As executives reduce the number of layers of management, the more matrix structures and their successor, cross-functional teams, become attractive. One problem with this sort of structure is that employees may find themselves reporting to two managers, and the allocation of authority between these managers must be addressed in organizations taking this approach. Because of this problem, many companies are finding cross-functional teams to be a more sensible approach. Figure 58 illustrates a typical matrix structure. Note that this structure facilitates bringing together the talents of many different people to work on projects.

FIGURE 58

In a matrix structure, functional managers are concerned with resource management. Cross-functional teams execute the processes by which work gets done.

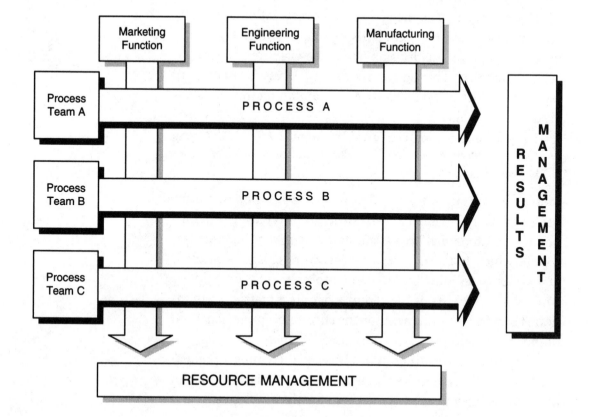

For more information, see: Charles Handy, *The Age of Unreason* (Boston: Harvard University Press, 1989); Peter F. Drucker, *Management: Tasks, Responsibilities, Practices* (New York: Harper & Row, 1973).

Mean This is the statistical term for the mathematical average of any group of data.

Mean Time Between Failures (MTBF) This is a measure of the average time between successive failures in the performance of a product or piece of equipment. It is one measure of reliability of product performance over time. The longer the mean time between failures, the higher the reliability of the product or machine.

For example, this measure can refer to the number of months between occurrences of a computer "going down" (failing to function). Improvements in design are intended to extend MTBFs from one generation of product to another.

For more information, see: N. Logothetis, *Managing For Total Quality: From Deming to Taguchi and SPC* (Englewood Cliffs, NJ: Prentice Hall, 1992); Armand V. Feigenbaum, *Total Quality Control*, Third Edition, Revised (New York: McGraw-Hill, 1991).

Mean Time Between Maintenance (MTBM) This is the average operating time of a machine between periods of maintenance to correct or prevent malfunction.

For more information, see: Armand V. Feigenbaum, *Total Quality Control*, Third Edition, Revised (New York: McGraw-Hill, 1991).

Mean Time to Repair (MTTR) This is the average time it takes to restore a system or machine to a specified state so it properly performs its required function. This, along with MTBM and MTBF, are all considerations when analyzing and evaluating how to reduce process cycle time.

For more information, see: Armand V. Feigenbaum, *Total Quality Control*, Third Edition, Revised (New York: McGraw-Hill, 1991).

Measurement More than just an important philosophical component of the quality practice's value system, it is an act or

Measurements are key. If you cannot measure it, you cannot control it. If you cannot control it, you cannot manage it. If you cannot manage it, you cannot improve it. It's as simple as that.

H. James Harrington

process of gathering quantitative data to compare actual results with intended requirements. Measurements of performance are nearly always expressed in quantitative terms. The idea of developing measurements is also called "metrics."

Measurements have expanded in scope beyond traditional costs or numbers of things. Today they include measures of how long tasks take to perform (cycle time), effectiveness (how well things work or how much they are appreciated by customers), and more sophisticated measures of efficiency (productivity of processes). The intent of this expanded use of measurements is to provide data for process control charts and, thus, a more quantitative, fact-based understanding of how processes are operating, and where opportunities for improvement lie. We can contrast this with more traditional reactive decision making based on single events and impressions rather than facts. The emphasis on measuring outcomes throughout processes and the development of new types of measurements lead to better understanding of system capabilities, earlier responses to process problems, and to deeper understanding of how to improve performance, productivity, and quality. Organizations often benchmark similar processes in other companies that are known to be "best in class" to provide a context for interpreting their measurements.

For more information, see: W. Edwards Deming, *The New Economics* (Cambridge, MA: MIT Center for Advanced Engineering Study, 1993); H. James Harrington, *Business Process Improvement* (New York: McGraw-Hill, 1991); George Stalk, Jr. and Thomas M. Hout, *Competing Against Time* (New York: The Free Press, 1990); Armand V. Feigenbaum, *Total Quality Control*, Third Edition, Revised (New York: McGraw-Hill, 1991); William Lareau, *American Samurai* (New York: Warner Books, 1991); Alexander Hiam, *Closing the Quality Gap* (Englewood Cliffs, NJ: Prentice Hall, 1992); Richard L. Lynch and Kelvin F. Cross, *Measure Up! Yardsticks for Continuous Improvement* (Cambridge, MA: Blackwell Business, 1991).

Median This is the middle value in a group of measurements, when arranged from lowest to highest. If the number of values is odd, it is the exact middle number. If the number of values is even, the convention is that the median is then the average of the two middle values. The median is helpful because it splits any set of values in half, with 50 percent being lower than this value and 50 percent being higher. The mean could be higher or lower than the median due to a few very high or low values affecting the average value of the entire group.

> For more information, see: Hy Pitt, *SPC for the Rest of Us* (Reading, MA: Addison-Wesley, 1994); and any other book that deals with basic statistical concepts.

Mentor This is an individual who takes on the responsibility to advise and guide a process team while serving as an interface between management and the team. Often a mentor is seen as a coach or quality advisor.

The concept of the mentor has taken on added significance in the past several years in two ways. First, the idea of using advisors to help process teams improve operations has been seen as a critical factor in team success. The use of mentors is now often standard practice in the creation of any organization's quality infrastructure. Mentors can either be full-time advisors to teams or serve in a part-time capacity; both approaches are effective. Second, as companies have reduced the number and layers of management, remaining managers have slowly acquired new roles that have been characterized as coaching, advising, cheerleading, facilitating, or mentoring employees, while dropping the role of "boss" of the past (also known as "command and control" practices).

> By teaching principles, sharing personal experiences, demonstrating good judgment, and pointing out acceptable and unacceptable behavior, a mentor can teach employees how to function and communicate inside a healthy company.
>
> *Robert H. Rosen*

> For more information, see: Jack D. Orsburn et al., *Self-Directed Work Teams* (Burr Ridge, IL: Irwin Professional Publishing, 1990); Abraham Zaleznik, *The Managerial Mystique: Restoring Leadership in Business* (New York: Harper & Row, 1989); Peter R. Scholtes, *The Team Handbook* (Madison, WI: Joiner Associates, 1988); Robert H. Rosen, *The Healthy Company* (Los Angeles: Jeremy P. Tarcher/Perigee, 1991).

Metrology This refers to the field of knowledge devoted to the study and use of metrics and measurements.

Micro-Process This describes a process executed by one department, team, or person.

The opposite is macro-process, which involves activities across multiple departments. A micro-process might be the activities done by an order entry clerk within a Customer Service Department, or the process used within a department to keep a pot of coffee constantly available to employees. (See *Macro-Process*.)

> For more information, see: William Lareau, *American Samurai* (New York: Warner Books, 1991); H. James Harrington, *Business Process Improvement* (New York: McGraw-Hill, 1991).

Military Standards The term refers to specific, documented standards of performance or quality issued by the U.S. government for any product employed by the military branches (also known as MIL-Q-9858A). Such standards encompass quality measures, performance criteria, statistical sampling reporting (MIL-STD-105E and MIL-STD-45662A), and specifications of components.

Since the U.S. military buys almost every kind of product made, their standards often influence design characteristics of commercially available goods.

> For more information, see: James H. Saylor, *TQM Field Manual* (New York: McGraw-Hill, 1992); Henry L. Lefevre, *Quality Service Pays: Six Keys to Success* (Milwaukee, WI: ASQC Quality Press, 1989); George L. Miller and LaRue L. Krumm, *The Whats, Whys & Hows of Quality Improvement* (Milwaukee, WI: ASQC Quality Press, 1992).

Minimum Acceptable Quality This is the maximum level of defectives or variants in a specified quantity of products, components, or services that, for the purposes of quality sampling, can be considered satisfactory as the average for the outputs delivered by a process. In other words, this term suggests a ceiling for the average number of defectives per lot and above which the lot becomes unacceptable.

> For more information, see: Joseph M. Juran and Frank M. Gryna, *Quality Planning and Analysis*, Third Edition (New York: McGraw-

Hill, 1993); Ellis R. Ott and Edward G. Schilling, *Process Quality Control*, Second Edition (New York: McGraw-Hill, 1990).

Mission This is the purpose of a process and what gives the actors in the process direction. For an organization, it is a statement of it purpose. It tells why the company exists and how it brings resources together to serve some particular segment of customers.

Managers have spent more time in crafting mission statements in recent years because of their need to be very precise in communicating the intent of the organization in a world in which empowered employees now make decisions without going through extensive management approval. In other words, as employees become responsible for deciding what they do, it becomes imperative that they understand the context in which to make their decisions. That requires a clearly stated mission, an sound strategic business plan, and measures in place to determine progress toward fulfillment of their mission.

Mission statements focus on how an organization uses its resources to serve some group of customers. In creating a mission statement, it is important to keep customers in mind and to focus on the benefits delivered rather than just on products. For example, a group of managers may define their company's mission as the development of a new type of computer. However, they need to tie that mission to benefits that this computer will deliver to some group of customers. Without this, they will not have a context in which to judge their success or give proportion to what should be invested in various efforts. It is important to note that a well-written mission statement will incorporate the connection between what the organization does and who it serves.

Managers often confuse mission statements with vision statements. An organization's mission describes its business and the benefits it delivers. A vision statement describes the direction an organization is headed and how it sees itself in the future. For example vision statements might include "Absolutely, positively, the best in the business!" or "We will change the

> Defining the purpose and mission of the business is difficult, painful, and risky. But it alone enables a business to set objectives, to develop strategies, to concentrate its resources, and to go to work. It alone enables a business to be managed for performance.
>
> *Peter F. Drucker*

world, one desk at a time." Here is an example of the mission statement of AT&T Universal Card: "We intend to be the primary and most helpful provider of our customers' transactions, payments, and selected services needs." Note that this matches up what the company does with the needs of customers.

In process improvement or reengineering activities, mission statements are seen as playing an important role because they give focus to teams. Studies of how best to manage process teams often point to the need for two early activities: (1) setting specific results-oriented expectations for the team, and (2) articulating the team's mission. Teams that do not perform these two tasks stand a greater chance of failing to be effective.

> For more information, see: Peter F. Drucker, *Management: Tasks, Responsibilities, Practices* (New York: Harper & Row, 1973); Burt Nanus, *Visionary Leadership* (San Francisco: Jossey-Bass, 1992); Jack D. Orsburn et al., *Self-Directed Work Teams* (Burr Ridge, IL: Irwin Professional Publishing, 1990).

Mistake Proofing A form of quality inspection, this is a technique to avoid human errors from occurring and thus affecting a process all the way down the line; fail-safing or poka-yoke are other names for mistake proofing.

Mistake proofing, or poka-yoke, is a weapon for avoiding simple human error at work. Because mistake proofing frees workers from concentrating on simple tasks and allows them more time for process improvement activities, it is a major weapon in the prevention of defects.

James H. Saylor

The goal is always to have zero defects. One conducts source inspections to find errors in a process before they can become defects. When a potential problem is found, you stop using the process until you find and eliminate the cause. For example, using this practice in a factory would mean that when an employee sees faulty outputs, he or she is authorized to shut down the line until the source of the faulty work is identified and fixed. In implementing mistake proofing, you have to monitor and measure processes at each step to catch problems early on, before they create defects. Built into an effective use of this practice are sound measurements, a reliance on statistical process control, and the empowerment of employees authorized to improve processes. (See also *Poka-Yoke*.)

For more information, see: Nikkan Kogyo Shimbun Ltd. (eds), *Poka-Yoke: Improving Product Quality by Preventing Defects* (Cambridge, MA: Productivity Press, 1988); Sarv Singh Soin, *Total Quality Control Essentials* (New York: McGraw-Hill, 1992); Joseph M. Juran, *Juran on Planning for Quality* (New York: The Free Press, 1988).

Mode This is the single value that occurs most frequently in a frequency distribution. See *Frequency Charts*, which graphically capture this value.

For more information, see: Hy Pitt, *SPC for the Rest of Us* (Reading, MA: Addison-Wesley, 1994).

Moment of Truth The phrase describes the moment when someone is face-to-face with a customer providing a product or service. These are always the points of contact between a customer and a vendor.

The phrase was originally coined at Scandinavian Airlines (SAS) to describe what happens when an employee sold a ticket to a customer. Now it refers to any customer contact: a telephone call, meeting, letter, and so forth. It embodies the idea that the quality of an organization's performance occurs and is measured at the moment when you deal with a customer and the customer with you.

In process improvement work, the voice of the customer is introduced frequently by looking at a flowchart of all the process steps and identifying at which points there is contact with a customer. Next, you describe the nature of that contact (such as phone call, in-person contact, or in writing) and determine how that contact works in terms of meeting and exceeding customer needs and expectations. Using that as a foundation and sometimes by benchmarking others, you can figure out what changes would represent improvements to enhance the quality of these customer-company encounters. The idea of moments-of-truth can also be helpful in reviewing and improving interaction among team members and between teams. By better understanding how these interactions affect performance, the company and its employees can make improvements that result in more smoothly operating and productive processes.

> For more information, see: Christian Gronroos, *Service Management and Marketing: Managing the Moments of Truth in Service Competition* (Lexington. MA: Lexington Books, 1990); Valarie A. Zeithaml, A. Parasuraman, and Leonard L. Berry, *Delivering Quality Service* (New York: The Free Press, 1990).

Multifunctional Team This kind of a team has members from various functions, usually brought together to work on a process that crosses more than two organizational functions. This is also referred to as a "cross-functional team."

The key idea here is that processes cross traditional functional lines within an organization. To improve a process, it becomes essential to look at its performance from beginning to end, regardless of who performs the process steps and in which departments. To do that requires two actions: (1) a team made up of representatives of various departments who do each step in a process, and (2) management commitment to operate the enterprise as a collection of processes, not simply as a cluster of functional departments. In implementing multifunctional teams, the company identifies interdependencies among departments and puts in place appropriate incentives to ensure cooperation.

> For more information, see: Jack D. Orsburn et al., *Self-Directed Work Teams* (Burr Ridge, IL: Irwin Professional Publishing, 1990); Bruce Brocka and M. Suzanne Brocka, *Quality Management* (Burr Ridge, IL: Irwin Professional Publishing, 1992); Alexander Hiam, *Closing the Quality Gap* (Englewood Cliffs, NJ: Prentice Hall, 1992); Peter R. Scholtes, *The Team Handbook* (Madison, WI: Joiner Associates, 1988).

Multivoting Used with brainstorming this is a process for selecting which items to work on from a list with limited discussion and difficulty.

Starting with a list of items generated by brainstorming, multivoting works like this:

1. Number all items on the list. If two or more ideas are similar, combine them, but only if all members agree.

2. All participants write down on a slip of paper the numbers of the items they prefer from the entire list. For example, if there are 50 ideas, each participant might write the numbers of 15 to 20 of these. Members should agree ahead of time on what amount of items each participant will select. Normally the number would be about one-third as many as on the entire list.

3. Tally the votes for each item on the list. In a group of eight participants, some items may have eight votes, some may have none or anywhere in between. After tallying the list, eliminate those with the fewest votes. For example, in a group of eight participants, eliminate those with three or fewer votes.

4. Go through the procedure again until the group has narrowed the list to a small number of items.

5. The leader then leads a discussion that will bring the group to consensus on their course of action, which may include some combination of the final items.

This procedure for voting on ideas is not meant to be rigid. Sometimes participants will not be able to come to a consensus decision in one meeting or even two meetings because they need to collect more information. However, they can still use this procedure to come up with a final set of ideas from which they can make their final consensus choice. (See also *Brainstorming*.)

For more information, see: Peter R. Scholtes, *The Team Handbook* (Madison, WI: Joiner Associates, 1988); William Lareau, *American Samurai* (New York: Warner Books, 1991).

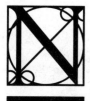

n This stands for sample size, such as the quantity of units in a sample from a lot. Often used in process work, a sample of n units will be selected from a lot to test the acceptability of the entire lot.

> For more information, see: Phillip J. Ross, *Taguchi Techniques for Quality Engineering* (New York: McGraw-Hill, 1988).

Need This is a requirement, often generated from what customers want, desire, or feel is essential. It is also used in design and manufacturing to define a lack of something that must be added for a process, function, or a product to be considered complete.

The notion of need is an important one in quality practices. The essential agreement among quality experts is that processes should deliver products or services that meet needs of others in the process and final customers. In satisfying customers, TQM recognizes that a company must do more than simply fill the articulated needs and expectations of customers. It must go beyond that to discover new ways of delivering products and services that exceed expectations and fulfill needs in ways that deliver completely unexpected benefits. This is what Deming and others call "customer delight." This idea is an old one: Thomas Edison may not have been told by customers that they wanted a lightbulb, but he knew that they could use such a product to give them what they did need: more light any time of the day or night.

You've chosen your customers wisely if: (1) The customers have needs you can meet and the means to make sure you're adequately paid for meeting them; (2) Your organization has, or can create, basic strengths that enable it to gain profitable market share by fulfilling those needs better than anyone else.

Richard D. Whiteley

For more information, see: W. Edwards Deming, *The New Economics* (Cambridge, MA: MIT Center for Advanced Engineering Study, 1993); Brian L. Joiner, *Fourth Generation Management* (New York: McGraw-Hill, 1994); Bob E. Hayes, *Measuring Customer Satisfaction* (Milwaukee, WI: ASQC Quality Press, 1992); Richard C. Whiteley, *The Customer-Driven Company* (Reading, MA: Addison-Wesley, 1991); Valarie A. Zeithaml, A. Parasuraman, and Leonard L. Berry, *Delivering Quality Service: Balancing Customer Perceptions and Expectations* (New York: The Free Press, 1990).

Nested Activity Center This is a term used in ABC accounting to identify an activity center within another activity center. Since each activity center has accounting and other measurements associated with it, by designating one activity center within another, you can more precisely assign costs while also noting how costs are related to each other.

Traditional accounting practices do the same thing. For example, sales statistics for a sales office can be considered those of a nested activity center, which are then rolled into those of the local sales district, which in turn are combined with those of other sales districts, to display the results in the performance of a sales vice president's organization. With the ABC approach, it is possible to measure the performance and costs of an activity or process at one level and accumulate them from others into a consolidated view. Thus, rather than accumulate data by functional organization (the traditional approach), you do it for activities or processes (which often go across functions)—the ABC approach.

For more information, see: Peter B.B. Turney, *Common Cents: The ABC Performance Breakthrough* (Portland, OR: Cost Technology, 1991).

Network(ing) This is a decentralized, often informal and undocumented organization within the official organization. It incorporates the unofficial communications channels and flow of information in any organization based on who knows who, friendships, rumors, and political alliances. Networks are a natural phenomenon in any organization as members seek to interact with each other on matters of mutual interest and concern.

In flattening organizations, where fewer managers provide formal channels of communication, and in which links between people are crucial, networking is an active and vital effort. Encouraging networks and informal communication among employees so they can find out what they need to know without going through the chain of command can make the organization run better. Indeed, technology gives everyone the ability to access information quickly and makes the chain of command often superfluous. E-mail now puts everyone in contact with everyone else in the enterprise. Fax machines allow for the quick and easy distribution of documents. Voice mail allows people to communicate by voice even when they do not speak directly to one another. Studies have shown that well over half the information someone receives in many organizations comes from networking activities. Rather than seeing this as bothersome, the new approach is to take advantage of it to make the organization run more effectively and efficiently. Indeed, one cause of inefficiency and consternation among employees is that managers try to buck or reduce networking and hoard information. In today's organizations, all this does is give employees one more thing to be concerned about (an information-hoarding manager) and replaces sound information that would be spread by the network with speculation and guesses.

> For more information, see: Charles Handy, *The Age of Unreason* (Boston: Harvard Business School Press, 1989); James Brian Quinn, *Intelligent Enterprise* (New York: The Free Press, 1992); Margaret J. Wheatley, *Leadership and the New Science* (San Francisco: Barrett-Koehler Publishers, 1992).

Noise Factors These are factors that are impossible to control in a process, experiment, or product usage that can affect variation in outputs or product performance.

Noise factors may include deterioration of parts or materials, customer usage conditions, temperature, voltage, humidity, and other such factors. Noise falls into three categories: inner, outer, and between product. Inner noise includes the tolerances between parts, specifications, and deterioration rates of parts and materials. Outer noise includes how the customer uses (or abuses) the product. This is beyond an engineer's control. Between product noise includes process-to-process variation and piece-to-piece variation. A goal in the design of products is to make them robust enough to withstand noise and perform properly.

> For more information, see: George L. Miller and LaRue L. Krumm, *The Whats, Whys & Hows of Quality Improvement* (Milwaukee, WI: ASQC Quality Press, 1992).

Nominal Group Technique (NGT)   This is a structured process to give everyone in a group an equal voice in helping the group decide on a course of action. It is especially useful in situations where some members might seem to dominate the group or when team members are new to each other. An advantage of NGT is that it allows a large number of issues to be pared down quickly. A potential disadvantage is that it discourages a lot of discussion. Here are the steps for NGT:

> Nominal group technique is an effective tool when all or some group members are new to each other. NGT is also good for highly controversial issues or when a team is stuck in disagreement.
>
> *Peter R. Scholtes*

1. Define the problem or opportunity so that everyone agrees on it and understands it. Write this on a flip chart in the front of the room.
2. Generate ideas. Each person individually and silently writes down on a sheet of paper all his or her ideas that address the problem or opportunity. There should be a set amount of time for this, approximately five to seven minutes. These ideas should be written as short phrases.
3. Record ideas on a flip chart. The leader goes around the room taking one idea from each person, continuing until all ideas are recorded. Number each item listed.
4. Discuss and clarify ideas. The leader goes through each idea, asking if anyone has questions or needs clarification. If possible, combine similar ideas.
5. Discuss and agree on criteria for voting on certain items. For example, the idea must be easy to implement,

acceptable to management, low cost, have a high likelihood of success, and similar criteria.

6. Rank vote the items. Depending on the number of ideas, the members are directed to select a smaller number of these and rank their preference for their selections. For example, if there are 25 ideas, each member selects 5. If there are 50 items, each person selects 8 to 10. The process for voting is as follows:

   • On three-by-five cards, each member writes down one selected idea in the middle of a card and the number of the idea chosen in the upper left corner of the card.

   • After selecting the agreed-on number of ideas, each member then ranks these from most preferred to least preferred. If there are eight items selected, the one *most* preferred is ranked as 8 and so on to the least preferred, which is ranked 1. The following figure shows what a card should look like:

> *12* (number of item)
> *Write phrase describing idea here*
> (rank of item) *4*

7. Collect the cards and tally the vote. The leader goes through and places the numbers showing the rankings for each idea. The final tally might look like this if there were 24 items:

1. 4-6-1	9. 6-1	17.
2. 3-8	10. 5-3-8-1	18. 7-2-3
3.	11.	19. 1-3
4.	12. 7-2-4	20.
5. 4-8-7-3-2	13.	21. 2-5
6.	14. 2-5-1	22.
7.	15.	23. 6-4-2
8.	16. 4-7-8-3	24.

7. Add the rankings for each item together, and the one with the highest total is the group's choice. In the list above, idea 5 has the highest total with 24. If the vote is close, select the two or three highest vote getters and through discussion come to a consensus about the idea that best meets the group's criteria for acceptance.

For more information, see: Peter R. Scholtes, *The Team Handbook* (Madison, WI: Joiner Associates, 1988); William Lareau, *American Samurai* (New York: Warner Books, 1991); Bruce Brocka and M. Suzanne Brocka, *Quality Management* (Burr Ridge, IL: Irwin Professional Publishing, 1992).

Nonconformity This suggests the lack of conformity to a pre-specified set of standards or customer requirements. It is often used as a synonym for defect, imperfection, or blemish.

It is also a code word in the world of quality, meaning any deviation from delivering exactly what a customer wants or any deviation from fitness for use. Philip B. Crosby, for example, speaks of conformity to requirements; W. Edwards Deming goes further to speak of conforming to customer needs. All writers on quality management address the importance of conforming to (and exceeding) customer needs and expectations and how to eliminate process problems that result in outputs that do not conform.

For more information, see: Philip B. Crosby, *Quality Is Free* (New York: McGraw-Hill, 1979); Joseph M. Juran, *Juran on Leadership for Quality* (New York: The Free Press, 1989).

Nondestructive Testing and Evaluation (NDE or NDT) This refers to testing and evaluation techniques that do not damage or destroy what is being tested, such as products or parts.

For more information, see: Armand V. Feigenbaum, *Total Quality Control*, Third Edition, Revised (New York: McGraw-Hill, 1991); Eugene R. Carrubba and Ronald D. Gordon, *Product Assurance Principles* (New York: McGraw-Hill, 1988).

Non-Value-Added This describes an activity, process, or task that adds costs and complexity but no value to the processes for delivering a product or service. An important part of process

> An organization can be vaccinated against nonconformance. It can be provided with antibodies that will prevent hassle. Some of these antibodies are managerial actions; some are procedural common sense. The organization that wishes to avoid internal hassle, eliminate nonconformance, save itself a bundle of money, and keep its customers happy must be vaccinated.
>
> *Philip B. Crosby*

improvement activities is finding non-value-added activities and changing the process to eliminate these.

Much non-value-added work comes from sloppy processes in which outputs do not pass inspection. This work involves fixing mistakes that might not happen if the process operated well. Many processes include a variety of steps to take care of contingencies. These include subprocesses to deal with faulty parts, missing parts, checking steps (for example, the need for multiple signatures to get something done), and many others. A goal of process improvement and reengineering is to find and eliminate these by changing the process or even throwing it out and setting up an entire new one. The overall purpose here is to root out those functions and activities that do not contribute to customer satisfaction.

> For more information, see: H. James Harrington, *Business Process Improvement* (New York: McGraw-Hill, 1991); James W. Cortada, *TQM for Sales and Marketing Management* (New York: McGraw-Hill, 1993); Brian L. Joiner, *Fourth Generation Management* (New York: McGraw-Hill, 1994).

Number of Defective Units Chart (np Chart) This kind of control chart is used to evaluate the stability of a process by looking at the total number of units with particular defects from a series of lots in which the sample size from each lot remains constant.

The plotted data are actual quantities (as opposed to percentages) of defects per sample (np). This chart might be used, for example, in determining billing problems. In a large company, a manager might examine 1,000 bills every four hours to detect errors of a certain type and note these on a control chart. As long as the number of errors remains within the control limits, the errors are due to common causes within the system. If the number of errors in any sample falls outside the upper or lower control limit, this indicates a special cause problem, which the company can immediately address. Figure 59 shows an np chart. It is not different from any other control chart except the sample size remains constant.

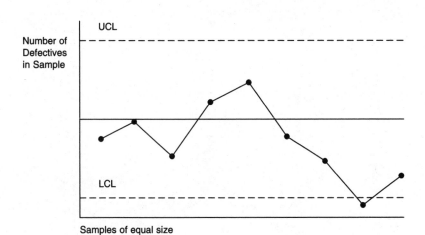

Number of
Defectives
in Sample

UCL

LCL

Samples of equal size

FIGURE 59

An np control chart.
What distinguishes this
as a control chart is that
the sample size remains
constant, and each point
on the chart indicates
the number of defectives
in that sample. In this
case, one sample
contains an amount
of defectives that falls
below the lower control
limit, indicating a
special cause problem
with that batch.

For more information, see: N. Logothetis, *Managing For Total Quality: From Deming to Taguchi and SPC* (Englewood Cliffs, NJ: Prentice Hall, 1992); George L. Miller and LaRue L. Krumm, *The Whats, Whys & Hows of Quality Improvement* (Milwaukee, WI: ASQC Quality Press, 1992); Hy Pitt, *SPC for the Rest of Us* (Reading, MA: Addison-Wesley, 1994).

Objective This is a desired, usually quantified, end result that a company, team, or individual wants to achieve within a specified period of time. Many people use the terms objective and goal interchangeably, and it is not productive to distinguish between the two.

In the world of process improvement and employee empowerment, teams and process owners set objectives, with the clear understanding of current process capabilities and what they might do to make improvements. They arrive at their objectives by taking into account customer requirements, competitive forces, and capabilities of existing or potential processes—information they can best derive by surveying customers and benchmarking the processes of other organizations. (See also *Goal.*)

For more information, see: Joseph M. Juran and Frank M. Gryna, *Quality Planning and Analysis,* Third Edition (New York: McGraw-Hill, 1993); Y.S. Chang, George Labovitz, and Victor Rosansky, *Making Quality Work* (New York: HarperBusiness, 1993); Christopher E. Bogan and Michael J. English, *Benchmarking for Best Practices* (New York: McGraw-Hill, 1994).

Operating Characteristic Curve (OC Curve) This kind of graph shows the probability of accepting a lot of goods compared to the percent of defectives in that lot.

The operating characteristic curve shows that the lower the percent of defects, the higher the chance of accepting a lot. It also demonstrates that there is some chance of accepting a lot of goods even if the number of defectives is higher than specifications call for (a bad lot). The goal is to use the graph to choose the sampling plan that will yield the lowest probability of accepting a bad lot. The OC curve measures sampling variation. It is not, however, a predictor of defects (as some have tried to use it) but rather a statement of the chances that a lot having a certain percent defects would be acceptable by a sampling plan. The shape of the curve for various plans is determined by statistical formulas and calculations. Figure 60 shows a typical operating characteristic curve for one type of sampling plan.

The important rule to follow is this: *Always determine the OC curve for any sampling plan used.* Otherwise you are sampling in the dark.

Hy Pitt

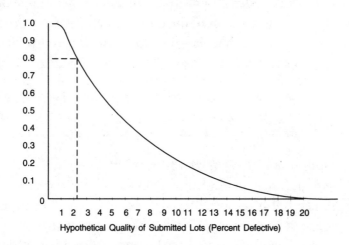

Probability of Accepting Lot Using Certain Sampling Plan

Hypothetical Quality of Submitted Lots (Percent Defective)

FIGURE 60

An operating curve showing that there is, for example, an 80% chance that a selected sampling plan will accept a lot with a little more than 2% defects in it.

For more information, see: Joseph M. Juran and Frank M. Gryna, *Quality Planning and Analysis,* Third Edition (New York: McGraw-Hill, 1993); Hy Pitt, *SPC for the Rest of Us* (Reading, MA: Addison-Wesley, 1994); George L. Miller and LaRue L. Krumm, *The Whats, Whys & Hows of Quality Improvement* (Milwaukee, WI: ASQC Quality Press, 1992).

Operational Definition This is the principle of defining a concept such that there is (1) agreement on the definition and (2) the definition is concrete in terms of action and can be used for business purposes.

Most practitioners also believe that operational definitions be specific to the point that they provide the basis for measurement. For example, the airline business defines "on time departure" as follows: "plane pushes away from the gate within 15 minutes of the scheduled departure time provided there is no aircraft mechanical difficulty." This definition is precise enough to allow for measurement—in this case, a count of the departures that meet this definition. Operational definition is an important concept in TQM. Deming devotes a chapter to it in his book *Out of the Crisis*, where he says: "Adjectives like good, reliable, uniform, round, tired, safe, unsafe, unemployed, have no communicable meaning until they are expressed in operational terms of sampling, test, and criterion" (pp. 276-277). Operational definitions are a key component of the documentation of any process. They eliminate confusion and misunderstanding among all parties.

For more information, see: W. Edwards Deming, *Out of the Crisis* (Cambridge, MA: MIT Center for Advanced Engineering Study, 1986); Peter R. Scholtes, *The Team Handbook* (Madison, WI: Joiner Associates, 1988); Gregory H. Watson, *Strategic Benchmarking* (New York: John Wiley & Sons, 1993).

Optimum   This is the state where the needs of customers and suppliers are met at minimal cost to both parties. In terms of practice, it is important to appreciate that in a dynamic, changing world, what is now optimum regularly changes based on new technology and changing customer needs and expectations.

Original Inspection   This is the first inspection of a batch or lot of goods. We contrast this with the idea of *reinspection* of the same lot if it had been rejected initially.

Out-of-Control Process   This is a process that, when measured, is delivering outputs that fall outside the bounds of normal variation or outside the control limits. These results are due to "special causes."

This concept is a critical idea in the measurement of process performance. It is one of the components in W. Edwards Deming's views on process variation. All processes will function within a certain range of performance, that is, within control limits. However, occasionally it will yield outputs that fall outside these limits. The reason for such events can almost always be traced to some special circumstances that have affected the process in an unusual manner, a special cause. The importance of understanding whether a process is in or out of control is that your techniques for dealing with process variation are distinctly different for in-control or out-of-control processes. A significant problem in many organizations is responding to variation that falls within control limits as if it were due to special causes. Doing this introduces new factors that tend to enhance rather than reduce the level of process variation.

> Stability is seldom a natural state [in a system]. It is an achievement, the result of eliminating special causes, one by one on statistical signal, leaving only random variation of a stable process.
>
> *W. Edwards Deming*

> For more information, see: W. Edwards Deming, *The New Economics* (Cambridge, MA: MIT Center for Advanced Engineering Study, 1993); W. Edwards Deming, *Out of the Crisis* (Cambridge, MA: MIT Center for Advanced Engineering Study, 1986); Brian L. Joiner, *Fourth Generation Management* (New York: McGraw-Hill, 1994); William Lareau, *American Samurai* (New York: Warner Books, 1991); William W. Scherkenbach, *Deming's Road to Continual Improvement* (Knoxville, TN: SPC Press, 1991).

Out of Spec This is a slang phrase that suggests something is not conforming to specifications, such as a part delivered by a supplier.

Output This is a term from general systems theory. It is a specific result or end product of a particular system process.

The systems view of organizations shows that work gets done in three stages: inputs, processes, and outputs. All members of an organization have the responsibility of working together to manage inputs and processes such that outputs will meet the quality expectations of customers. (See also *Input* and *System.*)

> For more information, see: W. Edwards Deming, *The New Economics* (Cambridge, MA: MIT Center for Advanced Engineering Study, 1993); Brian L. Joiner, *Fourth Generation Management* (New York: McGraw-Hill, 1994); Peter M. Senge, *The Fifth Discipline* (New

York: Doubleday Currency, 1990); Thomas H. Davenport, *Process Innovation: Reengineering Work Through Information Technology* (Boston: Harvard Business School Press, 1993).

Owner This refers to the person in an organization who has sufficient authority to change and improve a process without requiring approval. *Process owner* is a term often used to say the same thing.

Being responsible for the overall performance and improvement of a process includes:

- Process effectiveness (usually measured by number of defects).
- Process efficiency (usually measured by cost, including the cost of poor quality).
- Process control (documentation, measurement, clarification of duties, and improvement).
- Process adaptability (ability to deal with future constraints, deficiencies, and opportunities).

Process ownership usually suggests being the leader of the team responsible for executing the process. This could be a team in a single functional area or it could be a cross-functional team. In the best situations, ownership is taken on by all team members. They understand that the leader or formal owner serves as liaison with organizational management, making sure the team has the resources, training, and tools needed to perform well.

The concept of ownership indicates a shift away from functional organization and toward process management. This shift calls for managers to figure out how to allocate authority across the functional boundaries necessary to coordinate the actions by which high-quality products and services get delivered to customers. Finally, the idea of ownership also describes any employee's personal commitment to the company's values, culture, and way of operating. In that context, employees, as process owners, feel a responsibility to their team, the organization, and

On process ownership at Honeywell: A guy spent a whole bundle on an air heater and got it all done, and it was totally wrong—spent, I don't know, $10,000. But no one went back and said, "That's terrible, and you're getting your salary docked." We want to encourage that kind of thing. That's the epitome of the Honeywell story.

Dennis Bakke,
quoted in Manz and Sims,
Business Without Bosses

customers to maximize quality and minimize waste because they know that this benefits everyone.

For more information, see: Joseph M. Juran and Frank M. Gryna, *Quality Planning and Analysis*, Third Edition (New York: McGraw-Hill, 1993); Joseph M. Juran, *Juran on Leadership for Quality* (New York: The Free Press, 1989); Charles C. Manz and Henry P. Sims, Jr., *Business Without Bosses* (New York: John Wiley & Sons, 1993).

Parameter Design This is an approach to design in which an engineer identifies the numerical parameters for characteristics of inputs that will be most impervious to variation during the production or manufacturing process.

In any process, there are many possible causes of variation that are not completely controllable, often known as "noise" or "noise factors." Noise falls into three categories: inner, outer, and between product. Inner noise includes the tolerances between parts, specifications, and deterioration rates of parts and materials. Outer noise includes how the customer uses (or abuses) the product. This is beyond an engineer's control. Between product noise includes process-to-process variation and piece-to-piece variation. Understanding that these factors cause variation in the final quality of products, engineers and product designers can develop parameters for product components and assembly that minimize their effects.

This results in what is sometimes called *robust design*. This means the design can withstand many factors that might otherwise compromise the quality of the final product. Parameter design comes from the work of Genichi Taguchi, who also developed the Taguchi loss function (see entry *Loss Function*). Parameter design or robust design is a very important issue in quality management. Many statistical processes have been developed to facilitate this approach to design. Increasing design robustness is a central issue in process improvement and reengineering efforts. At the heart of many process measurement schemes is the concern for learning to what extent a process is robust.

> For more information, see: Joseph M. Juran and Frank M. Gryna, *Quality Planning and Analysis*, Third Edition (New York: McGraw-Hill, 1993); George L. Miller and LaRue L. Krumm, *The Whats, Whys & Hows of Quality Improvement* (Milwaukee, WI: ASQC Quality Press, 1992); Phillip J. Ross, *Taguchi Techniques for Quality Engineering* (New York: McGraw-Hill, 1988).

Pareto Chart Named after Vilfredo Pareto (1848-1923), it is based on his 80/20 rule. In quality management, a Pareto chart is a bar chart that illustrates in descending order the frequency of occurrence of particular events or process outputs.

The items documented on a Pareto chart might include accidents on the job or defects of different types. Pareto charts support an improvement strategy by helping to identify which types of problems occur most often. By dealing with these, you can have the greatest impact for the effort expended. For example, a manager might use this tool to identify which steps of a process yield the most errors. By gathering data on the process and then creating a Pareto diagram of the data, the manager can modify and simplify the tasks at those steps to reduce the possibility of errors. (See also *Leverage* and *80/20 rule*). Figure 61 (p. 252) illustrates a Pareto chart.

> For more information, see: Henry L. Lefevre, *Quality Service Pays: Six Keys to Success!* (Milwaukee, WI: ASQC Quality Press, 1989); Joseph M. Juran, *Managerial Breakthrough*, Revised Edition (New York: McGraw-Hill, 1994); Hy Pitt, *SPC for the Rest of Us* (Reading, MA: Addison-Wesley, 1994); George L. Miller and LaRue L. Krumm, *The Whats, Whys & Hows of Quality Improvement* (Milwaukee, WI: ASQC Quality Press, 1992), and many other books that deal with quality management tools.

Pareto Principle This widely known concept holds that a large percentage of results are caused by a small percentage of causes. Often called the 80/20 rule, it holds that approximately 80 percent of all problems are created by 20 percent of the possible causes.

FIGURE 61

A typical Pareto chart showing that most problems are of a similar nature or the result of one or two causes. In this figure dealing with the causes of problems 1 and 2 takes care of 72.5% of all problems.

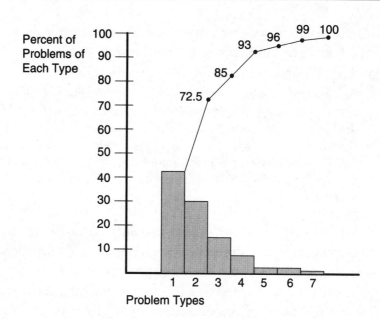

Seldom can we solve a complex problem all at once. Usually we need to focus in on specific components of the problem, which is what Pareto thinking and Pareto charts help us accomplish. Generally, we can get much greater leverage by focusing on reducing the largest source of trouble than we do from working to totally eliminate any of the smaller sources.

Brian L. Joiner

To use this principle effectively, you must quantify causes of problems or defects by the number of incidents and then rank them by order of frequency. This principle assumes that you can define the sources of problems correctly. The principle can be applied to all manner of activities. For example, if you are attempting to understand what are the most important irritants to customers and survey 100 customers, 49 may tell you poor telephone communication is the biggest issue, followed by 28 saying salespeople's lack of product knowledge, and another 23 having other miscellaneous complaints. From this you know that addressing the cause of poor telephone communications will improve customer relations the most. That is the Pareto principle at work. Proper management of processes calls for the use of the Pareto principle by use of a Pareto chart to document sources of problems. Fixing these sources of problems is one important way to improve continuously any process. Juran refers to the 80/20 rule as the "vital few" and the "trivial many." (See also *80/20 Rule* and *Leverage*).

For more information, see: Joseph M. Juran, *Managerial Breakthrough*, Revised Edition (New York: McGraw-Hill, 1994); Brian

L. Joiner, *Fourth Generation Management* (New York: McGraw-Hill, 1994); and many other books dealing with quality tools.

Participative Management In quality management, this term refers to employee involvement and authority over decisions and actions that cover the processes they implement or their individual work.

We often use the term in conjunction with the concept of employee ownership or empowerment. In this situation, management shares responsibility for the quality of process performance with employees or sets up self-managing teams and transfers responsibility for decisions and process management to them.

For more information, see: Jack D. Orsburn et al., *Self-Directed Work Teams* (Burr Ridge, IL: Irwin Professional Publishing, 1990); Charles C. Manz and Henry P. Sims, Jr., *Business Without Bosses* (New York: John Wiley & Sons, 1993); John O. Whitney, *The Trust Factor* (New York: McGraw-Hill, 1994).

People Involvement This describes the involvement of individuals or groups in managing processes and improving them. It is another variation of the concept of empowerment.

For more information, see: W. Edwards Deming, *The New Economics* (Cambridge, MA: MIT Center for Advanced Engineering Study, 1993); Jack D. Orsburn et al., *Self-Directed Work Teams* (Burr Ridge, IL: Irwin Professional Publishing, 1990); Charles C. Manz and Henry P. Sims, Jr., *Business Without Bosses* (New York: John Wiley & Sons, 1993); John O. Whitney, *The Trust Factor* (New York: McGraw-Hill, 1994).

Perceived Needs This refers to customers' wants based on what they think or believe they need. Perceived needs incorporate not only the physical benefits of a product or a service but also those having to do with self-image and all other aspects of the benefits delivered by an offering that a customer considers valuable. In understanding how customers make their purchasing decisions, perceived needs play the central role. Most organizations attempting to meet customer requirements begin by identifying and then satisfying perceived needs. The best of these companies then move on to anticipate customer needs based on a deep understanding of all the *benefits* their products or services might deliver to customers. (See also *Customer Delight.*)

For more information, see: Valarie A. Zeithaml, A. Parasuraman, and Leonard L. Berry, *Delivering Quality Service* (New York: The Free Press, 1990); Richard C. Whiteley, *The Customer-Driven Company* (Reading, MA: Addison-Wesley, 1991).

Percent Chart (*p* Chart) This is a commonly used attribute control chart on which all data is represented by percentages. A *p* chart is used to measure the proportion or percentage of non-conforming items from samples with the total number of units in each sample.

For example, you might use a *p* chart to track the percent of defective parts in 75 lots of 1000 units each of a component delivered by suppliers to a manufacturer by checking 100 sample units from each lot. Using standard statistical calculations, you can locate the upper and lower control limits of this chart, and determine the capabilities of the supplier's processes to deliver units that meet your specifications. The main difference between a *p* chart and a count chart (or a *c* chart) is that one uses percentages, and the other tracks the actual number of defects per sample. You calculate the percentage for each sample by dividing the number of defectives (d) by the total number of units in the sample (n): d/n = *p*. Figure 62 shows a *p* chart with the percentages shown on the y axis.

FIGURE 62

A *p* chart showing the percent defectives in 1,000 piece lots based on samples of 100 pieces from each lot. (Note: This only shows results from 13 samples, but most control charts should have at least 20 results to draw conclusions about the process.)

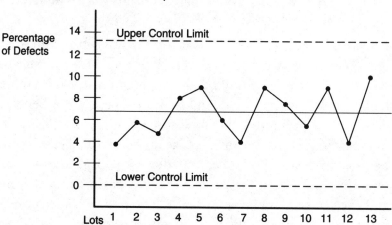

For more information, see: Hy Pitt, *SPC for the Rest of Us* (Reading, MA: Addison-Wesley, 1994); Henry L. Lefevre, *Quality Service Pays: Six Keys to Success!* (Milwaukee, WI: ASQC Quality Press, 1989); George L. Miller and LaRue L. Krumm, *The Whats, Whys & Hows of Quality Improvement* (Milwaukee, WI: ASQC Quality Press, 1992); and many other books on quality tools.

Performance This widely used word describes how the output of a process conforms to requirements and expectations, and suggests how well an individual, process, or team is operating.

The criteria most commonly associated with performance include costs (or process efficiency), quality (or effectiveness in meeting customer requirements), and availability or schedule (or cycle time). In process management, the best practice is to set targets for performance and then measure conformance against those targets. These measurements provide a quantitative way of judging performance and can serve as a basis for learning how to make improvements. TQM focuses on process performance and not on the performance of individuals, which almost always depends on the process they are part of.

For more information, see: H. James Harrington, *Business Process Improvement* (New York: McGraw-Hill, 1991); Bruce Brocka and M. Suzanne Brocka, *Quality Management* (Burr Ridge, IL: Irwin Professional Publishing, 1992); and many other books on quality improvement.

Phase System This refers to the means for breaking up a complex process into well-defined steps or phases.

The phase system defines work segments, establishes criteria each segment should meet, and provides for a management team to decide whether to continue or stop a process. For example, in the development of a complex software program, the team breaks the development process into a series of steps. At the end of each step, it can then evaluate progress, feasibility, and decide to proceed, modify the plan, or end it.

For more information, see: Joseph M. Juran, *Juran on Planning for Quality* (New York: The Free Press, 1988).

Physical Metrology This is that branch of metrology (measurements) that focuses on measuring mass, volume, density, pressure, and temperature.

Pilot Project This is a limited implementation of a new policy, procedure, process, product, or machine, under conditions of standard use. It may also be called a *pilot run* or a *pilot test.*

The purpose of pilot projects is to determine how well something performs and identify remaining problems and sources of defects and errors by a testing procedure. Pilots also help managers try out a new process or machine in one department or area before adopting it for the entire organization. Managers may initiate pilot projects to instruct an implementation team on how a process works, learn how they might improve it, and what they will have to do to install the full process. Best practices suggest that companies conduct pilot projects *prior* to the final decision to implement a new process or system. However, many times companies will make decisions to go ahead with a major project and use the pilot to learn the best way to proceed on the entire project.

By viewing pilots as an extension of more traditional experimentation projects, a company can be more aggressive in trying a variety of new ways of doing things, recognizing that not all will be successful or attractive. That approach avoids the pitfall of someone blaming a project team for disappointing results. It is a pitfall because such blame is seen as discouraging future experimentation—a by-product of poor pilot management.

For more information, see: Joseph M. Juran, *Juran on Leadership for Quality* (New York: The Free Press, 1989).

Plan-Do-Check-Act Cycle (PDCA Cycle) This is a basic approach to quality improvement in which you plan an action, do (or try) it, check to see how it conformed to plan and expectations, then make improvements to the process based on the check step, and then institutionalize the improvement. Then you do it all over again to make more improvements. This is often called either the

> In many companies the differences in viewpoints will preclude reaching a meeting of minds solely through discussion. In such companies, progress will depend on the results of a pilot test. If the results are favorable, the new approach is scaled up because the results are so attractive.
>
> *Joseph M. Juran*

Shewhart cycle or the Deming cycle, after the two individuals responsible for developing it and popularizing it. The idea of plan-do-check-act formalizes the process for continuous improvement. PDCA is often illustrated as a wheel going uphill, as shown in figure 20 (p. 80) under the entry *Continuous Improvement.*

> For more information, see: W. Edwards Deming, *Out of the Crisis* (Cambridge, MA: MIT Center for Advanced Engineering Study); Mary Walton, *The Deming Management Method* (New York: Perigee Books, 1986); Brian L. Joiner, *Fourth Generation Management* (New York: McGraw-Hill, 1994); Thomas H. Berry, *Managing the Total Quality Transformation* (New York: McGraw-Hill, 1991).

Poka-Yoke This term originated in Japan by eminent production engineer, Shigeo Shingo. It comes from two words: *poka,* meaning error, and *yokeru,* meaning to avoid. It refers to foolproofing a process.

There are a number of techniques and devices you can use in poka-yoke:

- Guide pins—to identify holes omitted in parts.
- Limit switches—to monitor proper placement and orientations.
- Checklists—to assure operator attention to procedure.
- Shut-off switches—to stop equipment when a machine senses an error condition.
- Counters—to make sure all parts have been used or actions completed.

These mechanical and memory devices and many others help operators prevent errors from occurring as they execute processes. The practice of poka-yoke also extends to the design of product parts. It suggests that parts should be asymmetrical so there is no confusion about how they go together. Figure 63 shows a simple example of applying poka-yoke to parts design.

> For more information, see: William Lareau, *American Samurai* (New York: Warner Books, 1991); Bruce Brocka and M. Suzanne

The intent of poka-yoke is to avoid, or immediately detect, errors caused by lapses of operator attention and vigilance by building safeguards into the system.

William Lareau

FIGURE 63

Part A is nearly symmetrical and subject to confusion when installing. part B is assymmetrical, making it impossible to install improperly. This is an example of poka-yoke design techniques. it helps mistake-proof an operation.

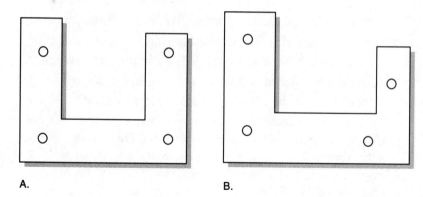

A.   B.

Brocka, *Quality Management* (Burr Ridge, IL: Irwin Professional Publishing, 1992).

Policy This is usually a documented statement designed to guide management and employee behavior in a variety of circumstances.

Policies are often used today to facilitate use of quality-based practices and values. They are important as general guides to what an organization intends and desires. When policies are properly conceived and articulated, management can more confidently implement empowerment and delegated decision making with the assurance that actions taken will be consistent with the policies.

This use of policies and teams makes it possible to reduce the number of managers required to run an organization, thereby reducing expenses and the time it takes to get decisions made and implemented. It is important not to confuse policies with rules and regulations. In many organizations, rules to deal with every possible contingency have been devised, and much energy can go into their enforcement. Experience suggests replacing such rules with more general policies, with the assumption that employees will not intentionally violate policy and will use it to make intelligent decisions. For example, a detailed set of company rules regarding funeral attendance of friends and relatives might be replaced with a general policy that says check with your supervisor and take the time you need. Companies that set such policies often discover that on average, employees take less time off than when there is a whole set of rules in place. Such policies, when intelligently administered, engender trust and teamwork rather than adversarial relationships.

For more information, see: Brian L. Joiner, *Fourth Generation Management* (New York: McGraw-Hill, 1994); John O. Whitney, *The Trust Factor* (New York: McGraw-Hill, 1994).

Policy Deployment This is a synonym for *hoshin planning* (see this term).

Population A population is a group of items with similar characteristics. We use statistical methods to study populations as an aid in decision making. A population is usually described within the parameters of space and time.

We can contrast population with the term *sample*, which suggests a portion of population. Though the term population may refer to people, it can also refer to a lot, a process, a batch, or any other word that suggests a grouping of items. Some examples of populations might include all television tubes manufactured at a single plant during the month of May, all inquiries made at a customer service facility during one week, all graduating seniors at the University of Wisconsin in June 1995, and so on. To understand the actual characteristics of a population, one would need to inspect or measure every item. Often this is neither possible nor practical. For this reason statisticians study samples of the population, which is a reliable alternative for learning about the whole group.

> The ability to make a distinction between a *population* and a *sample* will help assure that you have a clear notion of the items or data that you are studying, that you will be using the right language in expressing conclusions, and that you will be using the correct formula in certain instances.
>
> *Hy Pitt*

For more information, see: Hy Pitt, *SPC for the Rest of Us* (Reading, MA: Addison-Wesley, 1994); George L. Miller and LaRue L. Krumm, *The Whats, Whys & Hows of Quality* (Milwaukee, WI: ASQC Quality Press, 1992), and many other books that deal with basic statistics.

Potential Product This is any product or service that has not yet been developed that could be used to replace an existing product or service.

This offering has the *potential* to attract and retain customers beyond what existing products can do. Also known as "follow-on" or "replacement" products, once created they should always exceed performance of predecessor products or services. The key to potential products from the view of quality management is in ensuring that they meet or exceed customer expectations and that they perform better than predecessor products. During design engineering and in manufacturing, for example, potential products are subjected to rigorous testing and statistically based analysis for performance and feedback from customers.

> For more information, see: Roy A. Bauer, Emilio Collar, and Victor Tang, *The Silverlake Project* (New York: Oxford University Press, 1992); Thomas H. Berry, *Managing the Total Quality Transformation* (New York: McGraw-Hill, 1991).

Presentation Formally stated, this is the vehicle for providing data and obtaining approval for action. It is a formal way of discussing important issues. During presentations, managers may often use many of the charts and tools defined in this book to display data in graphic form to quickly make a point.

Process improvement teams consider reporting on the results of their work to be a critical, routine step in what they do. Simply having charts on the wall on process results is not enough. Effective teams take time to put together formal presentations on progress and results that can be critiqued by customers and by others affected by their work.

> For more information, see: James H. Saylor, *TQM Field Manual* (New York: McGraw-Hill, 1992); Peter R. Scholtes, *The Team Handbook* (Madison, WI: Joiner Associates, 1988).

Prevention In process management, this is an act taken upstream in a process to avoid defects or errors that manifest themselves downstream. It is a key idea in Total Quality Management. It focuses managers on dealing with the causes of problems rather than with symptoms.

Tied to the whole notion of eliminating errors and defects is the idea of avoiding or preventing them in the first place. We can do this by using statistical process control charts, cause-and-effect diagrams and analysis, and other quality tools that help operators identify and address the causes of process problems at their source. Deming, Juran, and Crosby all identify prevention as key to delivering quality outputs at the lowest cost. They recognize that because of variation in any process, there will always be problems, but the tools of quality allow for the measurement and analysis of variation and suggest what to address to reduce such variation and improve the quality of outputs.

In the past several years, the concept of the learning organization has developed, with the goal of everyone focusing on process and quality improvement and the prevention of problems. In these organizations, all processes are subjected to continuous inspection, documentation of past experience, and application of this knowledge to continuously improve operations and the quality of products and services.

> For more information, see: James Brian Quinn, *Intelligent Enterprise* (New York: The Free Press, 1992); Sarv Sigh Soin, *Total Quality Control Essentials* (New York: McGraw-Hill, 1992); W. Edwards Deming, *The New Economics* (Cambridge, MA: MIT Center for Advanced Engineering Study, 1993); Peter M. Senge, *The Fifth Discipline* (New York: Doubleday Currency, 1990).

Prevention Costs These are costs incurred in implementing actions to avoid future defects or errors in goods and services.

The idea is a simple one: to invest in the technology and training that will help the company prevent problems that occur during any type of process. Investing in prevention is almost always less costly than dealing with problems after the fact. For example, to avoid recalling hundreds of thousands of automobiles to replace a component that might fall off the car because it has only one bolt holding it on, you might go to the added expense of using two bolts. The cost of the second bolt and its installation on the car in our example might be $1.50 to hold a key part in place. Costs involved in replacing this part because it fell off are far higher than those involved in preventing that from happening

> If you look at [prevention costs] from a financial viewpoint, they are not costs. They are an investment in the future, often called a *cost-avoidance investment.*
>
> *H. James Harrington*

in the first place. Companies frequently incur prevention costs at the design stage of a product or service. This is the stage where such costs are at their minimum.

In practice, prevention costs occur in quality planning, assurance of components, dealing with vendors and subcontractors, reviews of designs and verifying their quality, design and development of quality-measuring devices, training, process control, quality auditing, use of process and quality data, quality improvement programs, planning and product recall strategies, and product liability insurance. (See also *Appraisal Costs* and *Poka-Yoke*.)

> For more information, see: H. James Harrington, *Business Process Improvement* (New York: McGraw-Hill, 1991); Armand V. Feigenbaum, *Total Quality Control*, Third Edition, Revised (New York: McGraw-Hill, 1991); Philip B. Crosby, *Quality Is Free* (New York: McGraw-Hill, 1979); William Lareau, *American Samurai* (New York: Warner Books, 1991).

Prevention vs. Detection This refers to the contrast between two approaches to quality-oriented actions. Prevention addresses actions taken to avoid defects while detection focuses on actions taken to find existing problems in goods and services. These two approaches are sometimes also said to be "designing in quality vs. inspecting in quality."

This concept is similar to prevention-based tactics in which you design in safeguards to avoid potential failures so long as the costs do not exceed the capability of customers to justify the expense of quality. Most quality experts would argue that designing in quality (prevention of defects) is more than cost justified because of lowered operating and manufacturing costs on the one hand and, on the other, customer satisfaction. Detection, when applied along with defect-avoidance designs, increases the opportunities of providing defect-free goods and services.

> For more information, see: Armand V. Feigenbaum, *Total Quality Control*, Third Edition, Revised (New York: McGraw-Hill, 1991); William Lareau, *American Samurai* (New York: Warner Books,

1991); Philip B. Crosby, *Quality Is Free* (New York: McGraw-Hill, 1979).

Primary Activities These are organizational tasks directly involved in the creation and delivery of a product or service to customers. These also incorporate customer service activities performed after the transfer of ownership.

Problem This refers to a circumstance or question that is a candidate for resolution. It is also the consequence of not conforming to a quality standard. The word is used frequently to describe a defect or error that needs to be resolved.

In traditional management, it is common to view an event or single data point as a problem, without having any understanding of the system as a whole or how well it is operating. For example, defects are 3 percent higher one week from the previous week. Managers may view this as a problem and take some action to correct it. However, it more likely represents some common cause variation in a manufacturing process. The proper way to understand this piece of data is in relation to many other similar pieces of data laid out on a control chart. In that way, managers can understand a distinct problem brought about a special cause from variation that is inherent in any process.

We may view excessive variation from an in-control process to be a problem if outputs regularly fall outside specification limits. The way you address that type of problem is quite different from how you address a special cause problem. Treating a common cause variation as if it were brought about by a special cause introduces the possibility of even more extremes in variation. In quality management, a primary assumption in dealing with problems is that they derive from the way a process operates, not from people implementing a process. Instead of trying to identify *who* caused a problem, you look at *what* aspects or relationships in the process caused it. The main use and value of quality tools and techniques is that they help us measure, understand, and fix the problems that occur in processes.

> Profits down by 10%? The almost immediate response is, "We want improvement now!" ("Strap on those guns and step out in the street pardner; it's time to be a hero.") The problem is that few people understand the system that caused the results. In the typical business environment, even fewer care; all they want is to come up with an answer, any answer, that will get the boss off their backs.
>
> *William Lareau*

For more information, see: Henry L. Lefevre, *Quality Service Pays: Six Keys to Success!* (Milwaukee, WI: ASQC Quality Press, 1989); Sarv Singh Soin, *Total Quality Control Essentials* (New York: McGraw-Hill, 1992); Mary Walton, *The Deming Management Method* (New York: Perigee Book, 1986); James H. Saylor, *TQM Field Manual* (New York: McGraw-Hill, 1992); Peter R. Scholtes, *The Team Handbook* (Madison, WI: Joiner Associates, 1988).

Process This is any activity or collection of activities that takes inputs, and transforms and adds value to them, and then delivers an output to an internal or external customer. A process has distinct start and end points and includes actions that are definable, repeatable, predictable, and measurable. A process always has a specific purpose that has value to some customer.

All actions are components of processes. A major problem in many organizations is the failure to recognize this fact. Instead they focus on individuals and events as if they were somehow isolated and not parts of a system and its processes. The successful management of processes to deliver outputs that customers will value is at the heart of TQM. Companies that move toward TQM spend considerable effort analyzing and documenting their processes, measuring their performance, and continuously working at making these processes more efficient and effective.

A well-managed process will display the following characteristics:

- *Clearly defined ownership.* The process owner is a manager who clearly owns the process in relation to the company's overall mission and knows what is necessary to accomplish that mission. There are standards by which performance is judged, such as schedule, quality, and cost. In many companies, the process owner is not an individual but a team.
- *Defined boundaries.* This means the beginning and end of the process is unambiguous and everyone understands what the inputs are and what outputs are expected.

- *Documented work flow.* Documentation provides the standard procedures for performing process steps in an acceptable and proven manner. It includes guidelines for training and serves as a baseline for making improvements.
- *Established control points.* These serve as a means for checking on and controlling the quality of work in the process. They are usually decision points on a flowchart and include such activities as inspection, verification, and disposition of nonconforming outputs.
- *Established measurements.* These are important for determining whether outputs at each stage of a process meet specifications and provide the data for process control charts. The type of measurements used are dependent on the process and product or service.
- *Control of process deviations.* Providing for timely corrective action of special cause variation based on SPC is an integral part of process management.

For more information, see: Peter R. Scholtes, *The Team Handbook* (Madison, WI: Joiner Associates, 1988); James W. Cortada and John A. Woods, *The Quality Yearbook 1995* (New York: McGraw-Hill, 1995); H. James Harrington, *Business Process Improvement* (New York: McGraw-Hill, 1991); Thomas H. Davenport, *Process Innovation* (Boston: Harvard Business School Press, 1993); William Lareau, *American Samurai* (New York: Warner Books, 1991); W. Edwards Deming, *Out of the Crisis* (Cambridge, MA: MIT Center for Advanced Engineering Study, 1986); Brian L. Joiner, *Fourth Generation Management* (New York: McGraw-Hill, 1994); Eugene H. Melan, *Process Management* (New York: McGraw-Hill, 1993).

Process Analysis Techniques (PAT) These are methods that define customer needs and analyze effectiveness of existing processes, leading to their improvement. They were most prominently defined by IBM and reflect common process improvement techniques. They are a variant of the PDCA cycle.

At the heart of PAT is the identification of steps in a process that add little or no value and steps that can be eliminated to simplify a process. In this effort, a critical measure of overall effectiveness is: Did a change in the process reduce its overall cycle time? In applying this technique, begin by analyzing a process and identifying nonvalue-added steps. Eliminate these. Measure the performance of the modified process against its old variant. Modify or standardize the new process. Start the cycle all over again.

For more information, see: James H. Saylor, *TQM Field Manual* (New York: McGraw-Hill, 1992).

Process Average This is a measure of primary tendency of a process to deliver outputs of specific characteristics. The process average is calculated like any average: Divide the sum of the measurement values by the number of items measured. Another way of defining the term is as the average percent of defects in a series of batches or groups.

For more information, see: Ellis R. Ott and Edward G. Schilling, *Process Quality Control*, Second Edition (New York: McGraw-Hill, 1990).

Process Capability This is a measure of the performance of any process as compared to its requirements or specifications for outputs. It is a measure of long-term performance after a process has stabilized (that is, it only takes in common causes of process variation). The process capability index (Cp) is normally how this is quantified. (See also *Cp.*)

The ability to statistically predict process capability makes this an important step in quality planning, especially in manufacturing. From a tactical point of view, a focus on process capability helps us predict the variability in a process compared to the specifications for final outputs. It helps us select from various alternative processes the one that will result in output with the least amount of variation from specifications. It also helps plan the sequence of steps in a process and determine the frequency and nature of process inspections. Finally, it serves as a way of testing various theories to learn why particular defects occur.

For more information, see: Joseph M. Juran and Frank M. Gryna, *Quality Planning and Analysis*, Third Edition (New York: McGraw-Hill, 1993); George L. Miller and LaRue L. Krumm, *The Whats, Whys & Hows of Quality Improvement* (Milwaukee, WI: ASQC Quality Press, 1992); Hy Pitt, *SPC for the Rest of Us* (Reading, MA: Addison-Wesley, 1994); Armand V. Feigenbaum, *Total Quality Control*, Third Edition, Revised (New York: McGraw-Hill, 1991).

Process Control This refers to actions taken that identify and eliminate special cause variation in a process, keeping the process operating within control limits.

It consists of three steps: (1) establishing points of control, (2) taking measurements, and (3) performing corrective actions. Points of control include steps in the process for checking, auditing, counting, or inspecting outputs. Measurements include: measures of conformance, measures of response time, measures of service level, measures of repetition, and measures of cost. Operators then take whatever action is appropriate depending on what the measurements indicate. Corrective actions are often specified as part of process documentation. An important goal of process control is the prevention or correction of sources of output problems as early in the process as possible. Process controls can be either automated or manual, depending on the nature of the process and costs. A thermostat is an example of an automated process control device.

For more information, see: Armand V. Feigenbaum, *Total Quality Control*, Third Edition, Revised (New York: McGraw-Hill, 1991); Eugene H. Melan, *Process Management* (New York: McGraw-Hill, 1993).

Process Design This refers to the specifics of determining how various resources—people, materials, and machines—come together to develop a process that prevents defects and minimizes variability in outputs. It also refers to the process of redesigning or reengineering an existing process.

Process design efforts tend to incorporate several common activities. First, people have a clear understanding of what is currently in place and what is wrong with it. This happens by applying a variety of analytical tools to establish facts of a process. Second, those responsible for the process should go through various brainstorming and team-building exercises to come up with ideas for design or redesign in response to pre-set requirements for results from a new process. Such requirements might include cycle time reduction, minimizing work-in-process, decreasing the number of people directly involved, decreasing the number of people indirectly involved (sign-offs required), reducing costs, and maximizing uptime of machines and people.

Juran and Gryna suggest that the design process should include such steps as:

- Creating a flowchart that breaks the process into its steps so you can see what is going on and what needs to happen.
- Identifying relationships between process variables and output results so as to prevent defects and minimize variability.
- Determining how to foolproof the process to minimize human errors.
- Deciding what employees need to know to implement their parts of the process correctly.
- Building in measurements and methods to implement SPC to keep the process operating properly.

Once the process is up and running, the company can compare (or benchmark) performance against that of other enterprises. To make this work best, a process team working with a process owner should have the authority to make decisions and implement the newly designed process.

For more information, see: Thomas H. Davenport, *Process Innovation* (Boston: Harvard Business School Press, 1993); Michael Hammer and James Champy, *Reengineering the Corporation* (New York: HarperBusiness, 1993); Peter R. Scholtes, *The Team Handbook* (Madison, WI: Joiner Associates, 1988); Joseph M. Juran and Frank M. Gryna, *Quality Planning and Analysis*, Third Edition (New York: McGraw-Hill, 1993).

Process Diagram This is another term for flowchart or other graphic representation of the various steps in a process.

This tool was borrowed from data processing where similar flowcharts have been in use for decades to describe each step that must be programmed. It is a useful way to capture visually all steps in a process, and it is the tool of choice in process documentation. When used in planning sessions, in which process definition, improvement, or just documentation are involved, the technique allows everyone to see how a process works and identify opportunities for improvement that come from eliminating complexity and redundancy.

> For more information, see: H. James Harrington, *Business Process Improvement* (New York: McGraw-Hill, 1991); Peter R. Scholtes, *The Team Handbook* (Madison, WI: Joiner Associates, 1988); James H. Saylor, *TQM Field Manual* (New York: McGraw-Hill, 1992).

Constructing flowcharts disciplines our thinking. Comparing a flowchart to the actual process activities will highlight the areas in which rules or policies are unclear or are even being violated.

H. James Harrington

Process Documentation This is the printed (or sometimes online) directions, including words and diagrams, people use to execute, manage, and improve a process.

An important facet of good process management is a very detailed and clearly laid out set of documents that describe the steps of a process as well as the information needed to perform each step, the inputs needed and specifications for outputs. It also includes a clear statement of expectations for performance (goals and targets) and a set of measurements for both the process operation and quality of final outputs. Good documentation practices also include keeping copies of earlier documents on how the process evolved, earlier measurements, and the results of any analysis of problems with the performance of a process. Process documentation is a requirement for process audits, many certification programs such as the ISO 9000, and applications for awards, such as the Baldrige or state quality awards.

> For more information, see: Peter R. Scholtes, *The Team Handbook* (Madison, WI: Joiner Associates, 1988); H. James Harrington, *Business Process Improvement* (New York: McGraw-Hill, 1991).

Process Drift This is a manufacturing term referring to variance in the tolerance of equipment, usually caused by wear and tear on the components that results in a process drifting slowly out of specification. Figure 64 illustrates this.

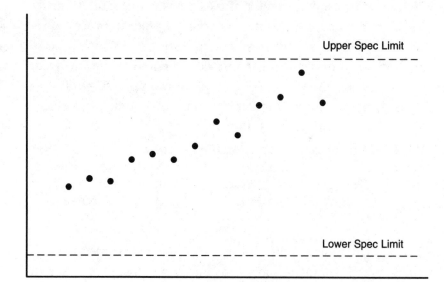

This problem occurs with such equipment as screw machines, drill presses, lathes, jig borers, and grinders. After a period of time, the tolerances and built-in controls of these machines can loosen up and become out of calibration, resulting in process drift. Drift can happen in any process where there is some laxity in applying SPC techniques, and performance varies over time away from original design specifications. Operators can usually only identify process drift by rigorously measuring process performance. Without such measures, drift can be occurring without the process users or owners realizing it. (See also *Trend*.)

> For more information, see: Armand V. Feigenbaum, *Total Quality Control*, Third Edition, Revised (New York: McGraw-Hill, 1991).

Process Efficiency Measurement This is a measurement of the value of outputs compared to resources, time, and money to create them. The higher the value of outputs compared to costs, the more efficient the process. One important measure of process efficiency is cycle time. Exploring cycle time at each step in a process can help to identify inefficiencies, bottlenecks, and steps devoted to rework or that result in wasted materials, all of which add costs but no value to outputs (thus compromising process efficiency).

For more information, see: H. James Harrington, *Business Process Improvement* (New York: McGraw-Hill, 1991).

Process Flow This is the series of steps and flow of materials through a process. It is usually depicted using a process flowchart.

For more information, see: H.G. Menon, *TQM in New Product Manufacturing* (New York: McGraw-Hill, 1992); H. James Harrington, *Business Process Improvement* (New York: McGraw-Hill, 1991).

Process Improvement This refers to the activities used to identify and then remove common causes of variation so as to improve the overall performance or capacity of a process. This is not to be confused with process reengineering which is the replacement of one process with a completely new and different one designed to deliver the same or better results more efficiently and effectively.

Two key concepts in quality management include the understanding that all work gets done as part of a process and that all processes can continuously be improved. From this understanding, experts have developed a set of tools and techniques for measuring process performance and identifying where to make changes that result in improved operations. These include root cause analysis, use of measurements, flow diagrams, PDCA, and several others documented in this book. All TQM efforts involve the continuous monitoring and improvement of processes.

For more information, see: Thomas H. Davenport, *Process Innovation* (Boston: Harvard Business School Press, 1993); Arthur R. Tenner and Irving J. DeToro, *Total Quality Management: Three Steps to Continuous Improvement* (Reading, MA: Addison-Wesley,

1992); James W. Cortada and John A Woods, *The Quality Year-book* (New York: McGraw-Hill, published annually).

Process Improvement Teams/Problem-Solving Teams These are temporary teams of people brought together with the specific purpose of solving a problem and/or improving a process. This type of team usually disbands after it solves the problem. However, such a team may become permanent if it comes to own the process.

> For more information, see: Peter R. Scholtes, *The Team Handbook* (Madison, WI: Joiner Associates, 1988); H. James Harrington, *Business Process Improvement* (New York: McGraw-Hill, 1991).

Process Management This is a collection of practices used to implement and improve quality management and process effectiveness within an enterprise.

Overall, managing a process includes defining, measuring, and continuously improving the way the process operates to satisfy customers. Eugene H. Melan notes the following seven aspects of process management:

1. Determine process ownership.
2. Delineate process boundaries and interfaces.
3. Define the set of work activities in a clearly understandable way.
4. Determine customer requirements.
5. Establish control points.
6. Measure and assess the process.
7. Obtain feedback and perform corrective action.

Process management frequently requires cutting across conventional organizational boundaries. It is not uncommon for a process to be maintained by a cross-functional team representing various departments and with a process owner as leader. They are authorized to make changes to the process and to command resources required to do so. Figure 65 shows the specific steps, activities, and tools involved in a process management as defined by AT&T.

The most important features of TQM are its customer orientation, its emphasis on continuous improvement, and its organization-wide aspects. Process management provides a means for addressing these features and services as a unifying methodology for TQM.

Eugene H. Melan

AT&T Process Quality Management and Improvement

Steps	Activities	Tools
1. Establish process management responsibilities.	Review owner selection criteria. Identify owner and process members. Establish/review responsibilities of owner and process members.	Nominal group technique
2. Define process and identify customer requirements.	Define process boundaries and major groups, outputs, and customers, inputs and suppliers, and subprocesses and flows. Conduct customer needs analysis. Define customer requirements and communicate your own requirements to suppliers.	Block diagram Survey Customer/supplier relations checklist Interview Benchmarking Affinity diagram
3. Define and establish measures.	Decide on effective measures. Review existing measures. Install new measures and reporting system. Establish customer satisfaction feedback system.	Brainstorming Nominal group technique Survey Interview
4. Assess conformance to customer requirements.	Collect and review data on process operations. Identify and remove causes of abnormal variation. Compare performance of stable process to requirements and determine chronic problem areas.	Control chart Interview Survey Pareto chart Cause-and-effect diagram Nominal group technique
5. Investigate process to identify improvement opportunities.	Gather data on process problems. Identify potential process problem areas to pursue. Gather data on subprocess problems. Identify potential subprocess problems to pursue.	Interview Flowcharting Brainstorming Pareto diagram Nominal group technique
6. Rank improvement opportunities and set objectives.	Review improvement opportunities. Establish priorities. Negotiate objectives. Decide on improvement projects	Pareto diagram Nominal group technique Trend chart
7. Improve process quality.	Develop action plan. Identify root causes. Test and implement solution. Follow through. Perform periodic process review.	Pareto diagram Nominal group technique Brainstorming Cause-and-effect diagram Force-field analysis Control chart Survey

FIGURE 65

A methodology for managing processes. Adapted from Joseph M. Juran and Frank M. Gryna, *Quality Planning and Analysis,* Third Edition (New York: McGraw-Hill, 1993).

For more information, see: Joseph M. Juran and Frank M. Gryna, *Quality Planning and Analysis*, Third Edition (New York: McGraw-Hill, 1993); H. James Harrington, *Business Process Improvement* (New York: McGraw-Hill, 1991); Eugene H. Melan, *Process Management* (New York: McGraw-Hill, 1993).

Process Map This is another term for a flowchart.

Process Optimization This is the act of improving both the efficiency and productivity of a process. In doing this, a process team will focus on economic issues and, more broadly, on productivity.

The basic strategy for process optimization is to carefully measure current performance at various points in the process and process results, all in quantitative terms. Next, you identify and root out special causes of malperformance while other steps are taken to improve efficiency and speed of performance. These other steps may include such tactics as eliminating redundant tasks, multiple audits, sign-offs, wait time between steps, and reduction in the number of steps within the process. You also set clear objectives for performance. Some strategies for optimization include a list of desired attributes in a process that are woven into the tasks. Examples include greater use of IT, more delegation of decision making to users, and less human labor to do the work.

For more information, see: Thomas H. Davenport, *Process Innovation* (Boston: Harvard Business School Press, 1993); H. James Harrington, *Business Process Improvement* (New York: McGraw-Hill, 1991).

Process Owner This is the person responsible for the implementation of a process and the person to whom management looks to manage its operation and improvement. This is also the person whom others in the process expect to interact with management to get the resources necessary to make sure the process operates properly. The process owner has authority to make decisions about the process without the approval of others.

Some things Juran suggests that an owner is responsible for include:

- Defining subprocesses.
- Ensuring the line-manager subprocess ownership is assigned and agreed upon.
- Identifying critical success factors and key dependencies to meet the needs of the business during the tactical and strategic time frame.
- Resolving or escalating cross-functional issues.

In addition, process owners establish measurements of success for a process, use continuous improvement tools and techniques to improve process efficiency and reliability, eliminate defects, and minimize variation. In organizations that understand the value of process owners, these individuals often head teams with members from various functional areas representing a recognition that many processes involve expertise from areas that are considered autonomous in traditional organizations. (See also *Owner.*)

> For more information, see: James H. Saylor, *TQM Field Manual* (New York: McGraw-Hill, 1992); Joseph M. Juran, *Juran on Leadership for Quality* (New York: The Free Press, 1989); H. James Harrington, *Business Process Improvement* (New York: McGraw-Hill, 1991).

Process Performance This generally refers to the effectiveness of a process. At a more formal level, it describes how well a process is operating to meet standards for efficiency in delivering outputs that meet specific standards, a selected benchmark, or customer requirements.

> For more information, see: H. James Harrington, *Business Process Improvement* (New York: McGraw-Hill, 1991).

Process Quality This refers to the ability of any process or subprocess to deliver outputs that meet design specifications. We measure process quality by employing statistical process control techniques.

The process owner must be able to anticipate business changes and their impact on the process. The owner must be at a high enough level to understand what direction new business will be taking and how it will impact the process.

H. James Harrington

For more information, see: J.R. Taylor, *Quality Control Systems* (New York: McGraw-Hill, 1989); Masaji Tajiri and Fumio Gotoh, *TPM Implementation* (New York: McGraw-Hill, 1992).

Process Quality Audit This is an analysis, appraisal, and evaluation of process performance against certain standards. The audit includes an evaluation of how operators maintain process quality and make accept/reject decisions about outputs.

For more information, see: Charles A. Mills, *The Quality Audit* (New York: McGraw-Hill, 1989).

Process Reengineering Often used interchangeably with process improvement, in the past several years it has become a distinctive term to describe those activities that lead to the creation of a new process radically different from its predecessor. Michael Hammer and James Champy, the management consultants who popularized reengineering, define it this way: "Reengineering is the fundamental rethinking and radical redesign of business processes to achieve dramatic improvements in critical, contemporary measures of performance, such as cost, quality, service, and speed."

Process reengineering is acquiring its own set of tools and techniques but retains the same criteria for performance sought after in process improvement, only on a more dramatic scale. It varies from continuous improvement in several ways. Unlike CI, organizations use reengineering only when an existing process must be changed in some radical fashion, for example, to achieve significant results beyond the capability of an existing process. Reengineering uses information technology as part of its base architecture, while in CI computers are almost an add-on. In CI team members typically make small changes and the process owner is only mildly involved; in reengineering the owner is intimately involved since the project will have a profound effect on organizational operations. Reengineering is also a one-time event while CI goes on for the life of the process.

For more information, see: Michael Hammer and James Champy, *Reengineering the Corporation* (New York: HarperBusiness, 1993);

> Reengineering promises no miracle cure. It offers no quick, simple, and painless fix. On the contrary, it entails difficult, strenuous work. It requires that people running companies and working in them change how they think as well as what they do.
>
> *Michael Hammer and James Champy*

James W. Cortada and John A. Woods, *The Quality Yearbook 1994* (New York: McGraw-Hill, published annually); Daniel Morris and Joel Brandon, *Reengineering Your Business* (New York: McGraw-Hill, 1993).

Process Review This is a formal, objective assessment of how well a process has been developed, performs, and is managed, not necessarily on the output of a process. The intent is to improve both the reviewed process and the methods used to improve and manage it. This is similar to a process audit.

Process reviews are properly done when the effort mimics the same steps used to audit accounting and financial activities. Someone other than the process owner conducts the review. This person documents the exercise and its findings, and the process team is required either to fix problems or look at and, if appropriate, implement recommendations for improvement. Like accounting audits, teams should conduct process reviews on a regular basis, against standards set by the enterprise for acceptable process performance. The value of such reviews lies in the information that comes from them. This can be used to help a process team make improvements.

For more information, see: John L. Hradesky, *Productivity and Quality Improvement* (New York: McGraw-Hill, 1988).

Process Routing Sheet This provides everyone involved in a process with a graphic depiction of each operation in the process. This is used mainly in manufacturing.

The process routing sheet provides everyone with the critical features of specific steps, the tooling involved with these steps, as well as the part names and numbers, customer name, operation number, and the date the sheet was set up along with all revision dates. This sheet is like an elaborate checksheet, customized for specific processes.

For more information, see: H.G. Menon, *TQM in New Product Manufacturing* (New York: McGraw-Hill, 1992).

Process Simplification This is the act of reducing the complexity of a process, usually by eliminating steps, such as redundant activities, multiple sign-offs, and steps that involve repair and rework. Removal of non-value-added activities is often the most effective way to improve the quality of a process.

> For more information, see: H. James Harrington, *Business Process Improvement* (New York: McGraw-Hill, 1991); Brian L. Joiner, *Fourth Generation Management* (New York: McGraw-Hill, 1994).

Process View This refers to how employees and managers understand the activities and structure of their organization. Rather than view themselves and their enterprise as a collection of functional departments and divisions, they see their work in terms of interrelationships and interdependencies that must be managed and improved to deliver outputs that satisfy customers.

Instead of organization charts, you see flowcharts of major processes and what resources are assigned to each. Sometimes process views are called functional views. In well-run companies, process maps are common. Associated with this view of business is a set of accounting reports and budgets more linked to activities than to departments (known as ABC accounting) in which measurements are of processes, not of departments or results. The argument put forth by quality experts essentially calls for organizations to take a process view of themselves and of their activities because only through this kind of perspective is it possible to improve performance in some continuous or dramatic fashion.

> For more information, see: James W. Cortada and John A. Woods, *The Quality Yearbook* (New York: McGraw-Hill, annual); William Lareau, *American Samurai* (New York: Warner Books, 1991); Thomas H. Berry, *Managing the Total Quality Transformation* (New York: McGraw-Hill, 1991).

Producer's Risk This is the risk of any producer that, through a sampling plan, a good lot will be rejected by a customer. This is different from consumer's risk, which suggests that a bad lot will be accepted.

> For more information, see: Joseph M. Juran and Frank M. Gryna, *Quality Planning and Analysis*, Third Edition (New York: McGraw-Hill, 1993).

Before deciding to adopt a TQM process, each company needs to address a very important question. Why bother? Specific answers to this question may vary from company to company, but generally the answer is this: BECAUSE IT'S A MATTER OF SURVIVAL!

Thomas H. Berry

Product This is the physical item that a company delivers for a customer to use or consume to fulfill some need. The key idea here is that people do not buy or value products; they buy the benefits they expect to derive from those products.

An important responsibility of any organization is to continuously improve its ability to meet and exceed customer expectations for the products it delivers. One way to do this is through quality function deployment, which is a tool and technique for matching up customer requirements with product characteristics. (See also *Augmented Product, Quality Function Deployment,* and *House of Quality.*)

> For more information, see: Sarv Singh Soin, *Total Quality Control Essentials* (New York: McGraw-Hill, 1992); Richard C. Whiteley, *The Customer-Driven Company* (Reading, MA: Addison-Wesley, 1991).

Product Activity This refers to all the procedures and processes associated with a product, such as design, design modifications, manufacturing, distribution, or marketing. The term is particularly important in companies using ABC accounting because the activities associated with products are assigned as costs as opposed to the more conventional approach by which costs are assigned to the product itself.

> For more information, see: Peter B.B. Turney, *Common Cents: The ABC Performance Breakthrough* (Portland, OR: Cost Technology, 1991).

Product Control Sometimes also called production control, this refers to all organizational activities involved in making sure products are defect free, fall within specification limits, and meet customer requirements. In effect, product control is the systematic application of statistical quality control tools and techniques to guarantee quality of final outputs to customers and prevent problems as early in a process as possible.

Armand Feigenbaum divides activities dealing with product control into seven steps:

1. Receipt of order for the part, material, or assembly in the manufacturing area.
2. Examination of the requirements of the order and taking the steps required to make the order ready for production, including product and process classification, correct assignment of all necessary equipment and controls.
3. Release of order to production.
4. Control of material during the manufacturing process.
5. Approval of product.
6. Quality audit with specific reference to safety and reliability considerations and evaluation of results.
7. Packaging and shipment.

For more information, see: Armand V. Feigenbaum, *Total Quality Control*, Third Edition, Revised (New York: McGraw-Hill, 1991).

Product Deficiency This refers to any product failure or shortcoming that results in customer dissatisfaction. This usually has to do with the customer finding the product fails to perform as expected. One way to judge quality is in terms of outputs being free of deficiencies or characteristics that would cause complaints by customers. Juran suggests that this can be represented in terms of an equation:

$$\text{Quality} = \frac{\text{frequency of deficiencies}}{\text{opportunities for deficiencies}}$$

The numerator can include such items as number of defective products, number of errors, number of failures with customers, and time and costs of rework. The denominator can include the number of units produced, total hours worked, total items sold, and sales dollars. The goal is always to make the numerator very small compared to the denominator. The closer to zero for the solution to this equation, the higher the quality.

For more information, see: Joseph M. Juran, *Juran on Leadership for Quality* (New York: The Free Press, 1989).

Product Design This term suggests all the activities involved in the planning and development of products that (1) meet and exceed the needs of customers, (2) can be delivered profitably, (3) are easy rather than difficult to manufacture, (4) are robust enough to withstand heavy use, and (5) are a better value and/or deliver more utility than the competition.

In quality management, product design is a rigorous set of processes that employ all the quality tools, ranging from statistical process control to customer surveys, flowcharting, quality function deployment (QFD), design of experiments, and poke-yoka. These techniques apply to products and services.

For more information, see: Sarv Singh Soin, *Total Quality Control Essentials* (New York: McGraw-Hill, 1992); Joseph M. Juran and Frank M. Gryna, *Quality Planning and Analysis,* Third Edition (New York: McGraw-Hill, 1993).

Product Dissatisfaction This refers to the negative effect on customers caused by deficiencies or performance failures in some product or service. Dissatisfaction often generates complaints, warranty work, returns, and other actions that are accounted for in the cost of poor quality.

For more information, see: Joseph M. Juran, *Juran on Leadership for Quality* (New York: The Free Press, 1989).

Product Feature This is any characteristic of a product that delivers some potential benefit to a customer.

Features are physical aspects of a product that make it easier to use, more functional, durable, aesthetically pleasing, and so on. Companies often focus on building additional features into their products without a good sense of whether customers will actually value the benefits delivered by these features. Therefore, consider carefully what customers will actually value before incorporating features that add costs but no additional quality as far as consumers are concerned.

For more information, see: Joseph M. Juran, *Juran on Leadership for Quality* (New York: The Free Press, 1989); Theodore Levitt, *Marketing for Business Growth* (New York: McGraw-Hill, 1974).

Product Genealogy This is a biological metaphor that refers to the historical record of the components and configuration of a product, and may include data about its manufacture, testing, inspections, maintenance, and the record of quality audits of both components and products.

> For more information, see: Armand V. Feigenbaum, *Total Quality Control*, Third Edition, Revised (New York: McGraw-Hill, 1991); Joseph M. Juran and Frank M. Gryna, *Quality Planning and Analysis*, Third Edition (New York: McGraw-Hill, 1993).

Product Goals These are quantifiable objectives that help define the characteristics of a product that will meet the needs and requirements of customers. These can include tolerances, hardness, dimensions, weight, costs—any characteristic that you can quantify and measure to determine if you have met your objectives.

> For more information, see: Joseph M. Juran, *Juran on Leadership for Quality* (New York: The Free Press, 1989).

Productivity Used widely with various meanings, it generally refers to the ratio of outputs produced or services rendered to the inputs required to deliver these goods or services.

Armand Feigenbaum defines productivity as "the effectiveness with which the resource inputs—of personnel, materials, machinery, information—in a plant are translated into customer-satisfaction-oriented production outputs and which today involve all the relevant marketing, engineering, production, and service activities of the plant and company rather than solely the activities of the factory workers, where traditional attention has been concentrated." This is a long sentence but it captures the relationship of productivity and satisfying customers. Without customer satisfaction, even the most efficient processes are not productive, but wasteful.

> For more information, see: Armand V. Feigenbaum, *Total Quality Control*, Third Edition, Revised (New York: McGraw-Hill, 1991); George Stalk, Jr. and Thomas M. Hout, *Competing Against Time* (New York: The Free Press, 1990); William Lareau, *American Samurai* (New York: Warner Books, 1991).

Generally, for every halving of cycle times and doubling of work-in-process turns, productivity increases 20 to 70 percent!

George Stalk, Jr. and Thomas M. Hout

Product or Service Liability This is the moral or legal obligation of a firm to make restitution to a customer for damage or personal injury caused by that company's products or services. This is often a source of cost justification for quality of products, especially for the cost of safety features. It is also part of the cost of quality.

> For more information, see: Joseph M. Juran and Frank M. Gryna, *Quality Planning and Analysis*, Third Edition (New York: McGraw-Hill, 1993); H. James Harrington, *Business Process Improvement* (New York: McGraw-Hill, 1991); Sarv Singh Soin, *Total Quality Control Essentials* (New York: McGraw-Hill, 1992).

Product Quality Analysis This is the analysis of what comprises quality in any particular product for specific market segments. Understanding these aspects allows for the intelligent deployment of policies and actions to deliver that quality.

Armand Feigenbaum lists a number of factors for a company to explore in undertaking such a product quality analysis. These include:

1. Customer-use needs and wants.
2. Functions of the product.
3. Environments in which the product may be used.
4. Durability and reliability requirements.
5. Regulatory and government industry requirements and standards.
6. Safety requirements.
7. Appearance and aesthetics.
8. Product design.
9. Manufacturability.
10. Shipping conditions.
11. Liability loss control.
12. Ease of installation
13. Maintenance and customer service.
14. Market characteristics.
15. Offerings by the competition.

A key factor in implementation of the quality system and in meeting the objectives and quality policy of the business is thorough analysis of the quality aspects of the product itself and those of the market served.

Armand V. Feigenbaum

For more information, see Armand V. Feigenbaum, *Total Quality Control*, Third Edition, Revised, (New York: McGraw-Hill, 1991).

Product Satisfaction This refers to those features of a product that evoke positive customer response. Customer satisfaction is not the opposite of customer dissatisfaction, which has to do with product deficiencies that elicit complaints. So eliminating defects and deficiencies in a product will not necessarily create satisfied customers. Satisfaction comes from a product delivering the benefits customers want and need and then even exceeding those to reach the level of delight.

For more information, see: Joseph M. Juran, *Juran on Leadership for Quality* (New York: The Free Press, 1989).

Proportion Chart This is another name for *percent chart* (see this entry).

Prosumers This is an invented term to signify customers who proactively participate in completing their own service or order fulfillment.

For example, if a consumer participates in the design of his or her own automobile by telling a dealer the color, engine, and so forth, this is a prosumer. In fact, it is in the design of automobiles that prosumerism first began to appear. Alvin Toffler originated the term to describe the more cooperative relations that would emerge between consumers and providers as they worked together to satisfy customer requirements.

For more information, see: Alvin Toffler, *The Third Wave* (New York: William Morrow, 1980).

Protocol This is a set of documented procedures or guidelines that govern the performance of individuals, products, or processes.

The phrase was originally used to describe how nations relate to each other and later, served as codes of conduct for organizations and individuals. In process management, protocols serve as a code of conduct for the purpose of standardizing sound practices. They are appearing most frequently in newly decentralized organizations with flattened hierarchies. Such protocols ensure the advantages of coordinated management without the burden of having a centralized management system.

For more information, see: James Brian Quinn, *Intelligent Enterprise* (New York: The Free Press, 1992).

Q9000 Series This is an abbreviation for ANSI/ASQC Q9000-1 series of standards, the U.S. version of ISO 9000 standards, adopted by the American National Standards Institute in 1987.

These are quality standards, documentation, and audit procedures for a variety of activities performed primarily by manufacturing organizations. Passage of these standards by an organization constitutes a form of quality certification similar to ISO 9000 certification.

> For more information, see: Charles A. Mills, *The Quality Audit* (New York: McGraw-Hill, 1989); John T. Rabbit and Peter A. Bergh, *The ISO 9000 Book*, Second Edition (New York: Amacom, 1994).

Quality While there are many definitions of the term, they all boil down to descriptions of excellence in goods and services and especially to what degree they conform to requirements and satisfy customers. Quality also includes freedom from defects and errors, thereby avoiding customer dissatisfaction. In this definition, it is equally important to understand that quality is a moving target and that processes, products, and services can be improved upon continuously. What seems high quality today can seem mediocre tomorrow. This is because people's expectations change based on what they learn about the strengths and weaknesses of any product and on what competitors offer. This is another reason companies need to work on continuously improving the quality and value of their offerings.

Armand Feigenbaum has captured the idea of quality in this way: "Quality is a customer determination, not an engineer's determination, not a marketing determination or a general management determination. It is based upon a customer's actual experience with the product or service, measured against his or her *requirements*—stated or unstated, conscious or merely sensed, technically operational or entirely subjective—and always representing a moving target in a competitive market." It is every organization's responsibility and mission to deliver quality outputs to its customers. The better companies do that, the more likely they are to generate the profit needed to continue in business. From this perspective, we can view profit as a measure of the quality of a company's products and services to its customers.

> For more information, see: Armand V. Feigenbaum, *Total Quality Control*, Third Edition, Revised (New York: McGraw-Hill, 1991); V. Daniel Hunt, *Quality In America* (Burr Ridge, IL: Irwin Professional Publishing, 1992); James W. Cortada, *TQM For Sales and Marketing Management* (New York: McGraw-Hill, 1993); Brian L. Joiner, *Fourth Generation Management* (New York: McGraw-Hill, 1994); William Lareau, *American Samurai* (New York: Warner Books, 1991); and several other books that deal with the basics of quality management.

Quality Audit This is a systematic inspection by an independent body to determine if quality activities conform to quality plans. A quality audit often reviews whether a company properly implements its quality plans and if these plans are appropriate for achieving the desired results. Such audits are a critical component of the ISO 9000 certification process and also a key step in judging applicants for the Baldrige and other quality awards. Quality audits are part of supplier certification programs carried by large companies such as Ford, GM, Chrysler, Boeing, Pillsbury, and many others. Charles Mills suggests that quality audits are used to determine one or more of the following about an organization and its processes:

World-class companies (and no company becomes world class without putting quality first) have achieved the following levels of performance in addition to the elimination of defects: 50% to 70% finished good inventory reduction; 70% to 90% work in process reduction; 40% to 70% space reduction; 30% to 50% capacity increase; 70% to 90% shorter lead times; 25% to 60% overhead reductions; 25% to 60% cost reduction.

William Lareau

1. The suitability of documentation as it applies to systems, products, services, processes, overall management practices, and so on.
2. The conformity or compliance of the operations to established documentation.
3. The effectiveness of the documentation and its implementation to achieve intended results.

In other words, a quality audit examines what a company says it will do and how, and compares that with actual practices and results. It then makes a judgment of the company's ability to consistently deliver quality outputs that will satisfy customers based on this audit.

For more information, see: Charles A. Mills, *The Quality Audit* (New York: McGraw-Hill, 1989); J.R. Taylor, *Quality Control Systems* (New York: McGraw-Hill, 1989); Mark Graham Brown, *Baldrige Award Winning Quality* (White Plains, NY: Quality Resources and Milwaukee, WI: ASQC Quality Press, 1991, new edition each year).

Quality Circle First made popular in the 1970s, this is a small group of employees and their managers who come together voluntarily to identify and solve work-related problems using various process management tools and techniques.

Using quality circles instead of TQM is like planting a few perfectly good seeds in unfertile soil. You'll get a few sprouts if you're lucky, but you'll never reap a profitable harvest.

Thomas H. Berry

For many American and European firms, quality circles represented their first attempt to implement quality management techniques. Quality circles did not usually work on process improvements, rather they focused on immediate problem solving. Typically members are from the same work area as the project or issue they are working on. Like self-managed work teams, however, they rely primarily on their own members for leadership in setting goals and determining what tasks to perform. The measure of effectiveness of quality circles is the number of completed tasks or projects, and the content of their solutions or improvements.

Early on companies thought that starting quality circles was tantamount to implementing quality management. However, because they did not do anything else and gave only superficial support to this activity, it has come into disrepute with employees and managers. In the 1990s, quality circles have become less popular than the use of process improvement teams as organizations and managers have come to realize the total change in values, culture, and methods that occurs when a company seriously adopts this approach to managing.

> For more information, see: Henry L. Lefevre, *Quality Service Pays: Six Keys to Success!* (Milwaukee: ASQC Quality Press, 1989); George L. Miller and LaRue L. Krumm, *The Whats, Whys & Hows of Quality Improvement* (Milwaukee, WI: ASQC Quality Press, 1992); Thomas H. Berry, *Managing the Total Quality Transformation* (New York: McGraw-Hill, 1991).

Quality Control This term refers to those activities a company and its employees undertake to ensure that organizational processes deliver high-quality products or services. In Japan, quality control is the term of choice for Total Quality Management.

Kaoru Ishikawa, one of the pioneers in Japanese quality management (and developer of the cause-and-effect diagram) describes quality control in this way: "Quality control consists of developing, designing, producing, marketing, and servicing products and services with optimum cost-effectiveness and usefulness, which customers will purchase with satisfaction. To achieve these aims. . . all the company's departments must strive to create cooperation-facilitating systems, and to prepare and implement standards faithfully. This can only be achieved through full use of a variety of techniques such as statistical and technical methods, standards and regulations, computer methods, automatic control, facility control, measurement control, operations research, industrial engineering, and market research." This statement, while repetitive of many others in this book (in terms of aligning the company processes with customer satisfaction), reinforces what is the heart of Japanese success. They believe in quality control and the use of quality management tools and techniques, and this explains their spectacular success in the commercial world.

> For more information, see: Kaoru Ishikawa, *Introduction to Quality Control* (Tokyo: 3A Corporation, 1990); Armand V. Feigenbaum, *Total Quality Control,* Third Edition, Revised (New York: McGraw-Hill, 1991); J.R. Taylor, *Quality Control Systems* (New York: McGraw-Hill, 1989); Joseph M. Juran and Frank M. Gryna, *Quality Planning and Analysis,* Third Edition (New York: McGraw-Hill, 1993).

Quality Council Typically, this is a committee of senior managers responsible for the oversight and implementation of quality management practices within an organization.

Quality councils have become very useful devices for promulgating and encouraging quality practices and a supportive corporate culture. Responsibilities of a quality council include oversight for quality education, strategies for quality transformation involving adoption of new corporate values, process views of the business, and audits of results. It is common for the chief executive officer to be a member of this council. His or her participation demonstrates dramatically the company's commitment to Total Quality Management practices and culture.

> For more information, see: James W. Cortada, *TQM for Sales and Marketing Management* (New York: McGraw-Hill, 1993); Thomas H. Berry, *Managing the Total Quality Transformation* (New York: McGraw-Hill, 1991).

Quality Engineering This incorporates the skills and expertise needed to apply statistical quality control techniques in the design and implementation of manufacturing processes to assure they operate efficiently, improve continuously, and deliver products that are free of defects and with minimum variation. Figure 66 shows the activities incorporated in quality engineering.

> For more information, see: Armand V. Feigenbaum, *Total Quality Control,* Third Edition, Revised (New York: McGraw-Hill, 1991); Joseph M. Juran, *Juran on Leadership for Quality* (New York: The Free Press, 1991).

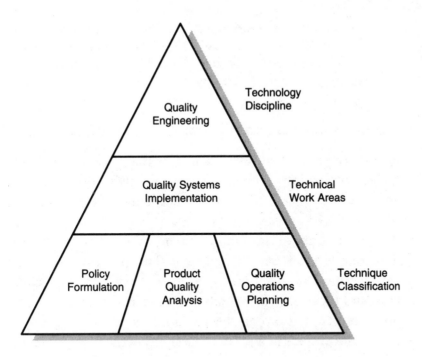

FIGURE 66

This represents the hierarchy of skills and activities in quality engineering. Adapted from Armand V. Feigenbaum, *Total Quality Control*, Third Edition, Revised (new York; McGraw-Hill, 1991), p. 236.

Quality Function Deployment (QFD) This is a highly structured methodology for identifying, classifying, and ranking customer requirements and expected benefits from a product or service, then correlating these to design features and production requirements. It is a disciplined approach to product planning and employs the *house of quality* to bring together several pieces of information.

In implementing QFD, design engineers, marketing, and manufacturing personnel work together to create a product design that most closely meets customer requirements. QFD includes the use of surveys to find out what benefits customers feel are important, followed by a ranking of features by significance, then by identification of possible problems, and concluding with well-defined engineering specifications. (See also *House of Quality* for figure.)

For more information, see: James L. Bossert, *Quality Function Deployment: A Practitioner's Approach* (Milwaukee, WI: ASQC

Another name for quality function deployment is *customer-driven* engineering because the voice of the customer is diffused throughout the product (or service) development life cycle.

Sarv Singh Soin

Quality Press, 1991); Sarv Singh Soin, *Total Quality Control Essentials* (New York: McGraw-Hill, 1992); Ronald G. Day, *Quality Function Deployment: Linking A Company with Its Customers* (Milwaukee, WI: ASQC Quality Press, 1993).

Quality Improvement This includes *all* organized activities to improve quality and value of products and services to customers.

Quality Improvement (QI) Team This is any team responsible for identifying improvement opportunities and then planning and implementing changes that enhance the efficiency and effectiveness of a process or processes.

QI teams are most useful in organizations that have only a moderate level of understanding of their processes and thus have significant opportunity to improve the efficiency, effectiveness, and focus of their operations. QI teams are usually composed of people from all the existing functional departments that perform steps in a given process. Properly managed teams have training in process improvement skills (such as statistical process control and problem analysis). They also have a process improvement coach or quality advisor assigned to help as needed. To be most effective, these teams have specific instructions as to what to accomplish (outcomes) and such other targets as time frames when results should be expected.

For more information, see: Peter R. Scholtes, *The Team Handbook* (Madison, WI: Joiner Associates, 1988); H. James Harrington, *Business Process Improvement* (New York: McGraw-Hill: 1991).

Quality Loss Function This is another name for the Taguchi loss function, developed by Genichi Taguchi. The loss function is represented graphically as a parabolic curve with the center being the target, and any movement away from the target representing a loss.

The quality loss function reminds managers that their mission is to make sure processes deliver outputs as close to target as possible. Any deviation from that represents a loss, a variation, and a compromise of quality, all of which raise costs. This approach contrasts with an approach in which any output that

> The Taguchi loss function recognizes the customer's desire to have products that are more consistent, and a producer's desire to make a low-cost product. The loss to society is composed of the costs incurred during the production process (for missing the target) and the costs encountered during use by the customers (repair, lost business, etc.).
>
> *Phillip J. Ross*

falls within certain specifications is acceptable. The quality loss function demonstrates that an important responsibility of managers is to reduce variation. The loss function is consistent with Deming's point that continuously reducing variation is the way to reduce costs while improving quality. Figure 67 illustrates the Taguchi loss function.

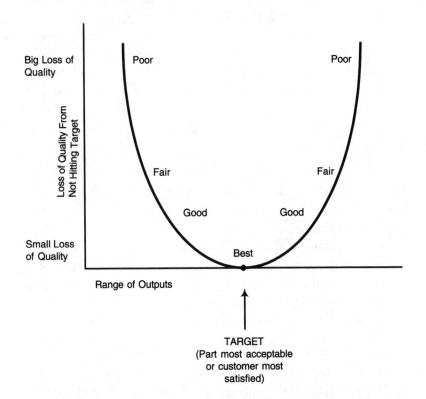

FIGURE 67

Quality (Taguchi) loss function. The farther from the target as specified by the customer or the specification, the more quality and value are compromised and the greater the possibility for waste and rework being introduced into the system.

For more information, see: Phillip J. Ross, *Taguchi Techniques for Quality Engineering* (New York: McGraw-Hill, 1988); Joseph M. Juran and Frank M. Gryna, *Quality Planning and Analysis*, Third Edition (New York: McGraw-Hill, 1993); Brian L. Joiner, *Fourth Generation Management* (New York: McGraw-Hill, 1994).

Quality Management This is the total of all techniques to manage quality in an organization, including products, services, performance, processes, and people. Quality management incorporates a quality trilogy common to all successful improvement programs: quality planning, quality control, and quality improvement, all implemented through teamwork.

> For more information, see: Joseph M. Juran and Frank M. Gryna, *Quality Planning and Analysis*, Third Edition (New York: McGraw-Hill, 1993); Joseph M. Juran, *Juran on Leadership for Quality* (New York: McGraw-Hill, 1989); W. Edwards Deming, *Out of the Crisis* (Cambridge, MA: MIT Center for Advanced Engineering Study, 1986); and many other books on TQM.

Quality of Working Life (QWL) This refers to the degree to which organizational culture makes available to employees information, knowledge, authority, responsibility, and rewards to conduct their work effectively and safely. It addresses the extent to which employees are compensated fairly and their personal dignity is respected. The premise for improving QWL is that a more healthy and supportive work environment brings out the best in employees. They feel a sense of loyalty, identification, and commitment to their company and work hard and intelligently to help it achieve its mission.

> For more information, see: Armand V. Feigenbaum, *Total Quality Control*, Third Edition, Revised (New York: McGraw-Hill, 1991); Hal F. Rosenbluth and Diane M. Peters, *The Customer Comes Second* (New York: William Morrow, 1992); Robert H. Rosen, *The Healthy Company* (Los Angeles: Tarcher/Perigee, 1991).

Quality-Planning Road Map This is defined by Joseph M. Juran as "a universal series of input-output steps, which collectively constitute quality planning." This suggests that the road map can serve as a template for the planning process in any organization. The stops on this road map are as follows:

1. Determine who your customers are, that is, what market segment you can serve best.
2. Identify what those customers need, that is, what benefits they are looking for or problems they have.
3. Relate those needs or problems into what you can deliver.
4. Design products and service features that will best respond to those needs.

5. Design processes that will allow you to efficiently and effectively produce, sell, and deliver these products and services.

6. Put the plan into operation throughout the organization.

Developing plans using this kind of road map requires the cooperation and participation of everyone affected by the plan. This suggests using cross-functional teams to work together in the planning process so that optimizing the operations of any one functional area, for example, marketing, does not compromise the operations of some other area, for example, manufacturing. This type of road map is consistent with *plan-do-check-act* (see this term), and at the end of the process you return to the beginning. Based on your learning, you continuously improve on how to use your resources to deliver quality and value to your customers.

For more information, see: Joseph M. Juran and Frank M. Gryna, *Quality Planning and Analysis*, Third Edition (New York: McGraw-Hill, 1993); Joseph M. Juran, *Juran on Leadership for Quality* (New York: The Free Press, 1989).

Quality Policy This is the set of statements that outlines an organization's overall intentions and objectives for the management and implementation of quality in managing processes to deliver products and services to its customer. This spells out the company's commitment to quality and what is acceptable and unacceptable. It is always a formal statement from senior management.

You can think of it as a kind of company credo that provides sound guidance for making decisions on process and product management and improvement. A quality policy, for example, may state that under no circumstances, even at financial loss, will the company ship deficient products to customers. Or it may include guidelines on supplier management, which empower employees to choose the supplier who can consistently meet their requirements, even if the prices are higher than another supplier. It can also include statements regarding the use of statistical process control tools and techniques, training,

> One of the most critical and difficult quality policy issues of a companywide nature is the need for early and thorough cooperation among the marketing, product development, and manufacturing functions ("tear down the walls").
>
> *Joseph M. Juran*
> *and Frank M. Gryna*

teamwork, and so on. The point is that such policies must be seen as gospel by top management. Then everyone will be committed to them. When quality is the number one priority within an organization, Juran and Gryna point out, then workers will not, for example, "succumb to schedule and cost pressures and classify product as acceptable that should be rejected."

The Japanese hoshin planning or policy deployment is a formalized procedure for aligning all levels and departments of the organization behind company quality policies and actions. It is a dynamic process that continuously upgrades what represents quality and in what ways the organization will use its resources to deliver quality.

> For more information, see: Joseph M. Juran and Frank M. Gryna, *Quality Planning and Analysis*, Third Edition (New York: McGraw-Hill, 1993); Armand V. Feigenbaum, *Total Quality Control*, Third Edition, Revised (New York: McGraw-Hill, 1991); William Lareau, *American Samurai* (New York: Warner Books, 1991).

Quality Standard This is any mandated quality performance criteria or requirements that must be adhered to in the manufacture of products or in the performance of services.

Quality System This is the body of practices, responsibilities, policies, and procedures used by an organization to implement and preserve levels of quality in products, processes, and services.

Using the term *system* with quality suggests that implementing quality requires an awareness that everything in an organization is interrelated and that it is only by managing the alignment of suppliers, internal processes, and distribution so as to create mutually beneficial relationships do you have a successful quality system. More generically, any program or set of procedures and actions to deliver outputs that meet customer requirements is a "quality system." For example, a quality program in a manufacturing plant designed along ISO 9000 standards is a quality system. A quality management approach based on the Baldrige Criteria is also a quality system.

For more information, see: Armand V. Feigenbaum, *Total Quality Control*, Third Edition, Revised (New York: McGraw-Hill, 1991); John T. Rabbit and Peter A. Bergh, *The ISO 9000 Book*, Second Edition (New York: Amacom, 1994).

Quality Team The phrase refers to quality improvement or performance action teams of employees, all working on designing, improving, or reengineering various aspects of a process.

Sometimes the phrase is used to designate a team of employees who have the responsibility to promulgate quality practices across an entire organization. Sometimes this team consists of the senior managers (frequently called Champions of Quality). In other organizations, this team may comprise the Manager of Quality or TQM and his or her staff, who provide training on quality practices, coach process improvement teams, and conduct quality assessments (for example, Baldrige-like reviews and process audits).

For more information, see: Thomas H. Berry, *Managing the Total Quality Transformation* (New York: McGraw-Hill, 1991); Alexander Hiam, *Closing the Quality Gap* (Englewood Cliffs, NJ: Prentice Hall, 1992); Peter R. Scholtes, *The Team Handbook* (Madison, WI: Joiner Associates, 1988); H. James Harrington, *Business Process Improvement* (New York: McGraw-Hill, 1991).

Queue Time This is the amount of time a partially finished product sits at one station in a process before someone works on it and passes it on to the next station.

In many processes, the actual amount of time spent working on a process output may represent around 5 percent of the entire amount of time it takes to deliver that output. The other 95 percent represents time it waits in queues for work to be done. An important goal in reducing cycle time, therefore, is to change processes so as to reduce queue time. One example is the manufacture of clothing. Most items of clothing are produced sequentially at stations with each station sewing one part of the garment, such as a hem or a sleeve. These parts are passed on to the next station where they sit in piles until the employee gets to them. These partially completed garments can sit around at each station for a week or more. Many clothing companies, such as Levi's, have changed this process by organizing people into teams who work together on entire garments. These teams pass partially completed pants or shirts among each other, with each having the flexibility to work on whatever part needs attention. The result is that garments are completed in hours rather than weeks.

> For more information, see: George Stalk, Jr., and Thomas M. Hout, *Competing Against Time* (New York: The Free Press, 1990); Gerard H. Gaynor, *Exploiting Cycle Time in Technology Management* (New York: McGraw-Hill, 1993).

Quincunx This tool creates frequency distributions, often used to simulate manufacturing processes. Steel marbles dropped through a funnel and fall through a series of nails arranged in groups of five (the source of the prefix "quin") to simulate sources of variability in a process. If the funnel is not shifted, the marbles will end up arranged approximately in a bell-shaped curve. Figure 68 shows a quincunx.

> For more information, see: Brian L. Joiner, *Fourth Generation Management* (New York: McGraw-Hill, 1994).

FIGURE 68

A quincunx, used to demonstrate variability in an in-control process. From Brian L. Joiner, *Fourth Generation Management* (New York: McGraw-Hill, 1994). Used with permission.

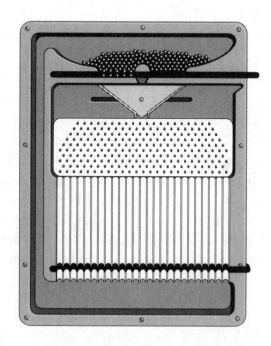

R Chart This is an abbreviation for *range chart.*

RAM or R&M This is an abbreviation for reliability/availability/maintainability. It is a process applied by the U.S. Department of Defense to ensure combat readiness of products through reliability and maintenance. This term is not to be confused with random access memory (also RAM), an information processing phrase.

The U.S. military quality process includes reduction of costs through defined practices. The U.S. Department of Defense has long been a leader in setting standards for products and components for its suppliers. Many of the statistical process control techniques used to measure quality of products and components were applied early to military contracts.

> For more information, see: James H. Saylor, *TQM Field Manual* (New York: McGraw-Hill, 1992); Armand V. Feigenbaum, *Total Quality Control*, Third Edition, Revised (New York: McGraw-Hill, 1991).

Random Sampling This is a standard sampling method by which random samples of units (for example, products) are chosen such that all combinations of these units have an equal chance of being chosen as the sample. This technique is used frequently for selecting products to be inspected for quality. It employs random number charts to decide which items from a lot will be chosen as samples.

> For more information, see: N. Logothetis, *Managing For Total Quality: From Deming to Taguchi and SPC* (Englewood Cliffs, NJ: Prentice Hall, 1992); Armand V. Feigenbaum, *Total Quality Control*, Third Edition, Revised (New York: McGraw-Hill, 1991).

Range This is the difference between high and low values of a set of data. Ranges are one way to measure the amount of variation in a process. Using the ranges or differences from one point to the next in a statistical control chart is how you calculate the standard deviation in a process and control limits. When calculating ranges, you always subtract the smaller number from the larger and ignore negative numbers. Ranges, thus, are always positive. For example, if a count chart yields the following data: 35, 28, 37, 32, 41, 38, 35, the range for this series would be: 7, 9, 5, 9, 3, 3.

> For more information, see: N. Logothetis, *Managing For Total Quality: From Deming to Taguchi and SPC* (Englewood Cliffs, NJ: Prentice Hall, 1992); John L. Hradesky, *Productivity and Quality Improvement* (New York: McGraw-Hill, 1988); Hy Pitt, *SPC for the Rest of Us* (Reading, MA: Addison-Wesley, 1994); and other books dealing with statistical process control methods.

Range Chart This type of control chart is a companion to the X-bar or average chart. It notes the range of values for any point on an average chart. The range chart helps us to discover if there is an increase in the variability of sample averages, which is not possible if we only note averages. In other words, while averages may not vary outside control limits, larger variation between the highest and lowest value contributing to any single point on the average chart may be increasing. This can indicate a special cause problem that would have gone undetected. Or it could indicate that the process, while in control, is operating less efficiently, increasing the possibility for outputs that fall outside specification limits.

Using the range chart, you create the center line by averaging the ranges. You create the upper and lower control limits on the range chart by multiplying the average range by a factor from a table. There are different factors depending on the number of units measured to find the average for each point on the X-bar chart. If the number of units is less than six, the lower control limit will be zero. This may mean that the average line of the ranges will not fall directly between the upper and lower control limits.

In creating average and range charts (and they are always used together), you first plot and create the control limits for the range chart. If you find that this indicates the process is out of control at any point, you can immediately address the special cause that is responsible. If you have started with the average chart, you might not have detected this out of control point because the averages would not have shown it. Figure 69 shows a range and average chart with plotted points (using the form from figure 5, page 30).

W. Edwards Deming has probably been the loudest voice in the past half century in articulating the benefits of using these kinds of tools to understand the performance of a process.

> For more information, see: W. Edwards Deming, *Out of the Crisis* (Cambridge, MA: MIT Center for Advanced Engineering Study, 1986); N. Logothetis, *Managing For Total Quality: From Deming to Taguchi and SPC* (Englewood Cliffs, NJ: Prentice Hall, 1992); Hy Pitt, *SPC for the Rest of Us* (Reading, MA: Addison-Wesley, 1994); George L. Miller and LaRue L. Krumm, *The Whats, Whys & Hows of Quality Improvement* (Milwaukee, WI: ASQC Quality Press, 1992).

Real Needs These are the basic or fundamental needs of customers that motivate them to buy a product or service.

A real need might be the need for transportation, not the need for an automobile; the need for medical attention is more fundamental than whether to obtain service from a hospital or a clinic. The significance of real versus perceived needs is that differentiating each makes it possible for a supplier to determine what customers will really pay for versus what they might pay for and, thus, affect the decisions on what features to provide to

> To understand our customers, we need to go beyond timely questionnaires. We must get inside our customers' lives, watching them use our product or service, discovering their aspirations and ways of life, their hopes and their fears. In that way we'll position ourselves to respond quickly, even anticipate critical needs long before they themselves recognize them.
>
> *Richard C. Whiteley*

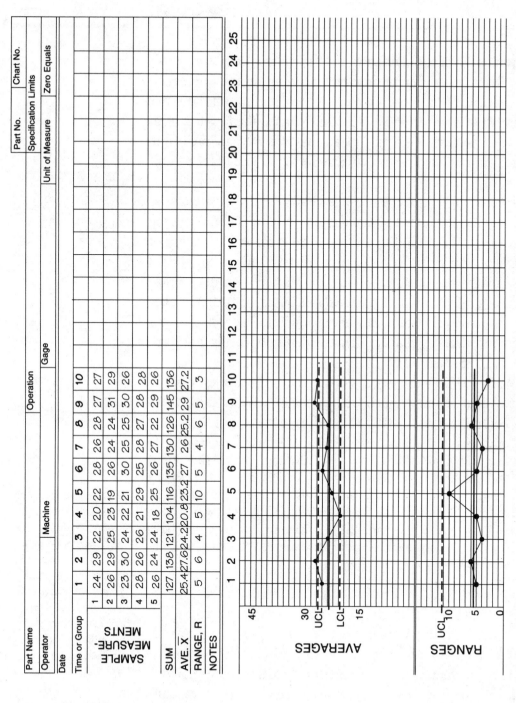

FIGURE 69 An average and range chart for ten groups of items. The average chart shows the process has special cause variation at various points. The range chart is in control.

customers. A real need refers to the actual benefit the customer receives as opposed to the product or particular service.

> For more information, see: Richard C. Whiteley, *The Customer-Driven Company* (Reading, MA: Addison-Wesley, 1991).

Real Time This term is borrowed from information processing. It means that a transaction or activity is done in the computer almost immediately. For example, looking up a piece of information in the computer using a terminal is a real time activity. There is very little wait time involved to receive the information.

The term has become more widely used outside of the IS community as people have become increasingly concerned about how to reduce cycle time in the performance of tasks within a process. It turns out that one of the most effective ways to speed up the work done within a process is to computerize some of the tasks, having them performed at the moment that a predecessor manual step has been completed (for example, data entry).

> For more information, see: Don Tapscott and Art Caston, *Paradigm Shift: The New Promise of Information Technology* (New York: McGraw-Hill, 1993); Gerard H. Gaynor, *Exploiting Cycle Time in Technology Management* (New York: McGraw-Hill, 1993); Thomas H. Davenport, *Process Innovation* (Boston: Harvard Business School Press, 1993).

Recognition This refers to the attention paid to either an individual or a group for outstanding accomplishments. In companies proactively practicing quality management, this is often tendered to teams. As a process, it is designed to acknowledge individuals for embracing and applying effectively quality tools and values. In developing a recognition program, it is important that this not undermine intrinsic motivation of team members. If it does, the goal of doing the job properly can be displaced by the goal of manipulating the system to gain some award or recognition.

> For more information, see: James W. Cortada, *TQM for Sales and Marketing Management* (New York: McGraw-Hill, 1993); James W. Cortada and John A. Woods, *The Quality Yearbook* (New York: Mc-

As Deming and others have emphasized, incentive programs reduce the possibility that people will cooperate. And when cooperation is absent, so is quality. "We talk about teamwork at training sessions," one bank executive remarked, "and we destroy it in the compensation system."

Alfie Kohn

Graw-Hill, published annually); Alfie Kohn, *No Contest: The Case Against Competition*, Revised Edition, (Boston: Houghton Mifflin, 1992); Alfie Kohn, *Punished by Rewards* (Boston: Houghton Mifflin, 1993).

Recovery This refers to actions of an organization to resolve an unanticipated problem, often brought about by a special cause. Recovery is called for when errors are committed by a company toward a customer or some uncontrollable situation occurs, such as bad weather.

Sound recovery strategies always focus on processes that address satisfying customers at the expense of the company. For example, customers are more likely to forgive an error if recovery is complete and done quickly (for example, replacing a broken product promptly with no questions asked and with apologies). Anticipating problems and having contingency plans to deal with them lead to quick and effective recovery (for example, restoration of electricity after a bad storm—you will have bad storms every year, so plan for them). Having the right information makes recovery possible (for example, having copies of all software and data file layouts at a location separate from your computer center for when the computer is flooded out). Also related to recovery is the set of practices known as risk management.

While it is important to prepare for contingencies and be able to recover quickly without losing customers, it is also useful to appreciate that such planning can also provide insights into what to do to minimize the need to recover. Problems often arise because of poor management processes, that is, from common causes rather than special causes. A weather problem is a special cause. Problems brought about from sloppy processes that allow defective products to go to customers are common cause situations. By reviewing processes and using TQM tools and techniques, you can reduce the possibility of problems, lower costs, and reduce the need for recovery.

For more information, see: G. Gordon Schulmeyer and James I. McManus, *Total Quality Management for Software* (New York:

Van Nostrand Reinhold, 1992); William Lareau, *American Samurai* (New York: Warner Books, 1991).

Red Bead Experiment This famous experiment was devised by W. Edwards Deming to teach a number of lessons about the performance of any system and the employees who are part of it. Deming included the red bead experiment as part of his famous four-day seminar.

The main gist of the experiment shows that poor and exceptional performance by individuals is the result of chance when a system is in statistical control. To rank or rate employees based on their performance becomes illogical and counterproductive when you look at work from this perspective.

The experiment is designed to simulate a production line. There is a container that has 4,000 small marbles (beads) in it, 20 percent of which are red. Using a paddle with 50 small holes into which the beads will fit, participants ("employees") dip into the container to come up with beads filling all the holes. The goal is to have only white beads, which the customer will buy. Any red beads represent defects. In performing this experiment, some employees have many red beads on their paddle. They are the "poor" performers. Some others get lucky and have only a few red beads. They are the "good" performers. The point is to show the absurdity of such judgments because the performance of both participants or employees was completely due to the way the system is set up.

Deming points out several lessons from this experiment that are worth noting as they form much of the basis for his approach to management. These include:

1. All variation in performance comes from the system itself. There is no evidence that one worker is better than another.
2. The experiment shows why the ranking of people as done in merit systems or performance appraisal is wrong and demoralizing. This is because the ranking is merely the

Having seen it once, Dr. Deming promises that you will never forget his simple experiment. "You will be seeing red beads wherever you go."

Mary Walton

effect of the process on people. Everyone is at the mercy of the system.

3. The experiment shows the futility of pay for performance. Performance is governed completely by the process employees work in.

4. Given the rules of the process, the workers had no opportunity to make suggestions or try new ways to improve the performance of the system. They each performed exactly as instructed, with no deviations. Thus there was no way the processes could be improved.

5. There was no evidence that the three best workers during any period would be the best workers in the future. This approach does not help managers make predictions about performance of individuals.

6. The managers themselves are as much victims of the system as anyone else, until they wake up and view work as processes in a system instead of individual excellence or mediocrity.

Deming added this line at the end of his lessons: "The reader may perceive Red Beads in his own company or in his own work." (p. 175, *The New Economics*). (See also *Funnel Experiment.*)

For more information, see: W. Edwards Deming, *The New Economics* (Cambridge, MA: MIT Center for Advanced Engineering Study, 1993); Mary Walton, *The Deming Management Method* (New York: Perigee, 1986).

Reengineering This term refers to any significant, often totally new, design of a product or process. Other items subject to reengineering include organizations, management systems, corporate values, and business strategies.

The purpose of reengineering is to respond effectively either to a requirement for significant improvements in performance or in response to fundamentally different circumstances. In the past few years, the approach of process reengineering has received considerable attention as a strategy for implementing quantum leaps forward in the improved performance of an organization at the process level. Such reengineering includes significant organizational changes, involving the restructuring of businesses with fewer layers of management, more empowered employees, and more extensive use of technology to reduce costs and speed up work. (See also *Process Reengineering.*)

> For more information, see: Michael Hammer and James Champy, *Reengineering the Corporation* (New York: HarperBusiness, 1993); James W. Cortada and John A. Woods, *The Quality Yearbook* (New York: McGraw-Hill, published annually); Daniel Morris and Joel Brandon, *Re-engineering Your Business* (New York: McGraw-Hill, 1993); Raymond L. Manganelli and Mark M. Klein, *The Reengineering Handbook* (New York: Amacom, 1994).

> [Reengineering] pursues multifaceted improvement goals, including quality, cost, flexibility, speed, accuracy, and customer satisfaction, *concurrently*, whereas the other programs focus on fewer goals or trade off among them.
>
> *Raymond L. Manganelli and Mark M. Klein*

Registrar Accreditation Board (RAB) This board evaluates the reliability and competence of organizations that assess and register companies to appropriate ISO 9000 series standards. This board was formed by the ASQC and is governed by members from industry, higher education, and management consulting.

> For more information, see: Bureau of Business Practice, *ISO 9000: Handbook of Quality Standards* (Milwaukee, WI: ASQC Quality Press, 1992).

Registration to Standards This is a defined process by which an accredited, independent third-party organization performs on-site audits of a company's operations against the set of requirements that the audited enterprise wishes to be registered. Used in certification for ISO 9000 standards, successful audits are rewarded with a certification of compliance to standards.

> For more information, see: John T. Rabbit and Peter A. Bergh, *The ISO 9000 Book*, Second Edition (New York: Amacom, 1994).

Regression Analysis This is a statistical technique for calculating the best mathematical expression that describes the functional relationship between a response and one or more independent variables. It is a widely used SPC calculation.

A regressor is an independent variable, usually called x, and is something we can control. A response is a dependent variable, usually called y, that can be observed at different levels of *x.* Regression analysis allows one to estimate the functional relationship between *x* and *y*, leading to the prediction of the values of *y.* Scatter diagrams are frequently used in graphically noting data for regression analysis. The mathematics, while not too difficult, are somewhat involved, and it requires a trained person to properly undertake this analysis.

This tool is used when quality problems need to be studied to understand the relationship between two or more variables. Companies effectively use it for forecasting, locating optimum operating conditions, prediction, and the importance of some variable in influencing others.

> For more information, see: Joseph M. Juran and Frank M. Gryna, *Quality Planning and Analysis,* Third Edition (New York: Mc-Graw-Hill, 1993); N. Logothetis, *Managing For Total Quality: From Deming to Taguchi and SPC* (Englewood Cliffs, NJ: Prentice Hall, 1992).

Reliability This is the study of the probability that a product or service will perform as intended under predetermined conditions and for a planned period of time (the longer the better, generally).

Reliability is an important aspect of quality as defined by customers. Companies make their products more valuable to customers by making sure they are highly reliable and will perform as expected over a long period of time. Reliability includes four aspects that are of interest to engineers (and customers). These are: (1) probability, which deals with how likely it is that a product will operate properly within a given time span; (2) performance, which deals with how well the product performs its assigned function for an employee or a customer; (3) time, which is about the length of time a product remains functional; (4) conditions, which deal with what kind of environmental considerations are required for a product to perform well.

Quality engineers have developed a number of measures for product reliability. These include:

- *Time to Wear Out.* This has to do with the longevity of the product.

If product reliability is too low, actual total costs to the customer may be high because of excessive repair, maintenance, and out-of-use costs. If an unduly high reliability level is evolved, total cost may still be excessive to the customer because of the higher price caused by unique requirements for components and assemblies.

Armand V. Feigenbaum

- *Time to Uniform Replacement.* This has to do with the longevity of various components in a product. If repairs of components keep a product functioning, then it does not have to be replaced. This measures when it is more appropriate to replace a product than repair its components.
- *Mean Time Between Replacement.* This is a measure of the average time before a product must be replaced.
- *Mean Time Between Maintenance.* This measures how long a product can operate before it requires routine maintenance to prevent malfunction.
- *Mean Time to Repair.* This measures how long a product will be out of service during a repair (is it quick or time consuming to repair).

There are a variety of tests that have been developed to determine the reliability of products or processes. The main tests include:

- *Environmental Stress Screening.* This helps to point out weaknesses in new designs and failure in the product because of weak parts or defects in processes or other environment-related causes of defects or nonconformance.
- *Reliability Development Tests.* Companies undertake these tests before releasing a product to production. They usually involve a protocol to discover problems that can be solved before going to full production.
- *Reliability Qualification Tests.* Companies conduct this test on parts that are representative of an actual production run.
- *Reliability Acceptance Tests.* Companies use these tests to check production equipment to make sure no step in the production process will introduce defects. It is often undertaken after retooling or other changes have been made.

For more information, see: Armand V. Feigenbaum, *Total Quality Control,* Third Edition, Revised (New York: McGraw-Hill, 1991); Joseph M. Juran and Frank M. Gryna, *Quality Planning and Analysis,* Third Edition (New York: McGraw-Hill, 1993); H.G. Menon,

TQM in New Product Manufacturing (New York: McGraw-Hill, 1992); Dev G. Raheja, *Assurance Technologies* (New York: McGraw-Hill, 1991).

Reliability Engineering This is the branch of engineering devoted to improving product performance. It includes a set of practices that focus on accurately predicting when and under what circumstances products or processes might fail or not deliver acceptable outputs. Using that information, companies can improve product designs and set operating limits for equipment. They also work with others to develop fail-safe and backup procedures and build redundancies into a product if this is cost effective and/or demanded by customers. Best practices include procedures for delivering feedback to design engineers working on product improvements.

For more information, see: J.R. Taylor, *Quality Control Systems* (New York: McGraw-Hill, 1989); Dev G. Raheja, *Assurance Technologies* (New York: McGraw-Hill, 1991).

Remanufacturing This is the overhauling of goods returned to a company after having been used by customers.

In this process, a remanufacturing company purchases used products (for example, motors of some type), repairs and refurbishes them, using new parts as necessary, and then resells them. For example, for years computers have been refurbished by second party leasing companies. The machines are repainted, computer chips are replaced, and then the machines are leased to new customers. With engines, and manufacturing equipment of all types, this can involve normal repairs, replacement of electronic components with more effective and efficient ones, and repainting. More than an ecological move, remanufacturing is a new form of manufacturing that is gaining in popularity and requires its own set of processes.

For more information, see: Richard J. Schonberger, *Building a Chain of Customers* (New York: The Free Press, 1990); Jack Stack, *The Great Game of Business* (New York: Doubleday Currency, 1992).

Remedial Journey Along with the "diagnostic journey," this is a term coined by Joseph M. Juran to discuss the cycle of events that begins with an understanding of known causes of problems in a product or process and ends with an effective resolution. It is a basic quality-improvement process to help define team activities by which a company can enhance its performance.

> For more information, see: Joseph M. Juran, *Juran on Leadership for Quality* (New York: The Free Press, 1989).

Representative Sampling This is a process by which samples are pulled from batches or lots of units so as to contain minimum bias between the values of the samples' characteristics and the batch or lot as a whole. In other words, by testing and finding that 10 percent of representative samples are unacceptable, you can reliably extrapolate that finding to be true for the entire lot.

> For more information, see: Ellis R. Ott and Edward G. Schilling, *Process Quality Control*, Second Edition (New York: McGraw-Hill, 1990); Hy Pitt, *SPC for the Rest of Us* (Reading, MA: Addison-Wesley, 1994).

Requirements These are always captured in a formal statement of customers needs and expectations. Requirements also note expected or unstated *minimum* levels of performance, such as, that your airplane will be operated safely, that the automobile will conform to government safety regulations, that food you buy is safe. Requirements are an important influencing factor in determining the content and output of any process. In specifying requirements, the best practice is to include measurements we can use to know whether a product or service meets customer requirements. Companies should also develop plans for achieving these requirements.

As organizations improve the performance of their processes, they move along a continuum beginning with guessing what customers want next, to specifically knowing their requirements, to accurately anticipating their future needs, and finally to customer delight. Companies can come to understand customer requirements today and in the future by studying con-

Remember that customers are a moving target. We need to stay close to them if we are to serve them well. A Delighter one year quickly becomes a Must Be the next: TV remote controls were Delighters a few years back, but now are Must Be's for most people.

Brian L. Joiner

FIGURE 70

A continuum demon-
strating levels of under-
standing of customer re-
quirements.

sumer buying habits, appreciating environmental conditions that might influence consumer buying (such as changes in the law, tastes, and costs of raw materials), and through an understanding of what product enhancements would be needed once a product was in wide use. Customer delight comes from carefully examining exactly how products are used, what benefits they deliver, and then making it exceptionally easy to attain those benefits or adding unexpected benefits that customers will value. Figure 70 shows a continuum of learning for successfully meeting customer requirements. (See also *Customer Delight.*)

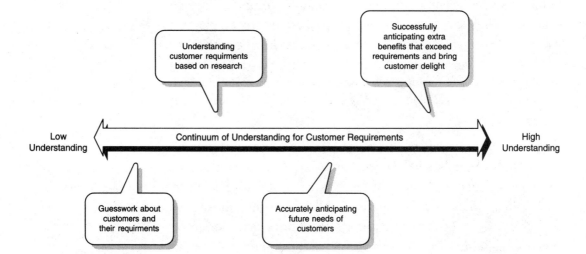

Requirements also change. Yesterday's "out-and-above" or "delighter" becomes tomorrow's minimum requirement. For example, when automatic teller machines (ATMs) first appeared, a bank could offer 24-hour banking and enjoyed a competitive advantage. Customers were delighted with a new service they had not demanded in the past. Today all banks have ATMs and could not compete without them. The appearance of these machines changed customer expectations for banking services. Finding new ways to deliver services and new services to deliver that will surprise and attract customers is thus an on-going challenge. This is yet another driver of the need to continuously improve processes, products, and services.

> For more information, see: H. James Harrington, *Business Process Improvement* (New York: McGraw-Hill, 1991); Brian L. Joiner, *Fourth Generation Management* (New York: McGraw-Hill, 1994); Bruce Brocka and M. Suzanne Brocka, *Quality Management* (Burr Ridge, IL: Irwin Professional Publishing, 1992); Valarie A. Zeithaml, A. Parasuraman, and Leonard L. Berry, *Delivering Quality Service* (New York: The Free Press, 1990).

Resource Driver These are the links between resources and activities. In ABC accounting systems, costs flow from the general ledger to an activity by the resource driver.

For example, suppose there are two resources used with a particular activity (such as, $50,000 in salary and $5,000 in supplies). In ABC systems, those two amounts would be assigned to a particular activity as the cost of that activity. The estimation of what effort is required to perform the activity leading to the assignment of cost is known as the resource driver.

> For more information, see: Peter B.B. Turney, *Common Cents: The ABC Performance Breakthrough* (Portland, OR: Cost Technology, 1991).

Resources These are the available assets an organization can deploy in the performance of the processes it uses to deliver value to customers. People, money, equipment, buildings, and experience are all examples of resources.

As companies form intimate bonds with suppliers, stakeholders, and customers, their definition of the types of resources available broadens. But, even in these instances, resources are assets of economic value deployed in the execution of a company's activities. Accounting practices, both traditional and ABC, still catalog only assets that are legally the property or in the employment of the company. However, process managers and their teams are coming to view all stakeholders who participate in process execution as part of the resources available to help achieve the company's mission. Much of Total Quality Management is about getting better and better at using resources to profitably deliver quality to customers.

> For more information, see: Peter B.B. Turney, *Common Cents: The ABC Performance Breakthrough* (Portland, OR: Cost Technology, 1991); H. James Harrington, *Business Process Improvement* (New York: McGraw-Hill, 1991); Brian L. Joiner, *Fourth Generation Management* (New York: McGraw-Hill, 1994).

Restraining Forces These are realities that prevent a situation from changing or improving. They can be intrinsic to a process (for example, culture, lack of funds, lack of experience) or be caused by some circumstance outside of the organization (for example, geographic location, government regulation).

The term is frequently used in force field analysis in figuring out what is involved in making a change. You compare restraining forces with driving forces to learn how to balance off the two to achieve some goal for improvement. (See also *Force Field Analysis.*)

> For more information, see: James H. Saylor, *TQM Field Manual* (New York: McGraw-Hill, 1992); Bruce Brocka and M. Suzanne Brocka, *Quality Management* (Burr Ridge, IL: Irwin Professional Publishing, 1992).

Retrospective Analysis This is any analysis based on information from prior operations or earlier experiences.

This type of analysis goes with Juran's concept of "lessons learned" where managers can take note of what has happened over time, observing cycles or patterns of influence and causes by which events occur. By using this kind of analysis, you can learn where to intervene to make changes that will have the greatest leverage or the possibility of breakthrough to higher levels of performance.

> For more information, see: Joseph M. Juran, *Juran on Leadership for Quality* (New York: The Free Press, 1989).

Reverse Engineering This is the process of comparing one product to a competitor's function-by-function, performance-by-performance, and price-for-price, then the tearing down (breaking apart) of the competitor's products to their sub-assembly levels for another comparison and, finally, the manufacture of similar products for the same market.

Reverse engineering has long been a strategy for coming up with products that compete toe-to-toe with another's relying on reproducing as closely as legally allowable a new product. In benchmarking, it is a formal process for learning about a competitor's product, its strategy, its costs, its processes, and its service capability. All of this can be valuable for understanding where a company stands in comparison to its competitors and what its products must be like to appeal to chosen market segments.

For more information, see: Christopher E. Bogan and Michael J. English, *Benchmarking for Best Practices* (New York: McGraw-Hill, 1994).

Reward This is recognition for something well done. In quality management, there is a strong effort to distinguish between external and internal rewards. External rewards are given to individuals or teams by management or peers for quality results. Internal rewards are the personal satisfaction of meeting a challenge, doing a job well, or accomplishing a personal goal.

Two of W. Edwards Deming's 14 Points indirectly address this issue of internal and external rewards: Points 11a and b, "Eliminate numerical quotas for the workforce and for management." Point 12, "Remove barriers that rob people of pride of workmanship." Numerical goals set people toward "making their numbers" by whatever means to gain an external reward, which can limit performance and distort the way the system operates. Further externalizing rewards and taking away from people the opportunity to make complete use of their skills in order to keep their jobs and receive some reward evokes compliance rather than commitment to the system.

It is not yet clear what the best reward and recognition programs are. However, we understand that people like to know they are doing a good job. This reward can affirm their own sense of what needs to be done and provide direction for future

> As behaviorists cheerfully admit, theories about rewards and various practical programs of behavior modification are mostly based on work with rats and pigeons.
>
> *Alfie Kohn*

behavior. However, when it is done to manipulate performance and behavior, it can lose its effectiveness as a way of improving individual and company performance. (See also *Recognition* and *Motivation*).

> For more information, see: James W. Cortada, *TQM For Sales and Marketing Management* (New York: McGraw-Hill, 1993); Alfie Kohn, *No Contest: The Case Against Competition*, Revised Edition (Boston: Houghton Mifflin, 1992); Alfie Kohn, *Punished by Rewards* (Boston: Houghton Mifflin, 1993).

Right Sizing This term came into wide use in the early 1990s to describe changing the size of the workforce in large corporations.

In some companies, it has become a euphemism for "downsizing" or reducing a company's workforce en masse. However, for organizations practicing Total Quality Management, the phrase implies the idea that there is a right size for any company. This size is the one where labor and other resources are not wasted doing rework and solving quality problems. This happens because the company has adopted tools and techniques that continuously reduce such problems. If a company downsizes without also changing its processes, this can simply compound problems because there are now even fewer people to operate the company's already inefficient and ineffective processes.

> According to a study published by *U.S. News & World Report*, a typical company's stock price rose 10% after downsizing. Three years later, the stock had dropped by 35%.
>
> *Stephen George and Arnold Weimerskirch*

> For more information, see: Stephen George and Arnold Weimerskirch, *Total Quality Management* (New York: John Wiley & Sons, 1994).

Right the First Time People use this phrase to emphasize prevention of problems and to suggest that it is less expensive and more beneficial to make a product or deliver a service correctly the first time. Taking steps to ensure that products and services perform as required is seen as more cost beneficial than having to "do it over again." In other words, it is another way of suggesting that defect prevention strategies are more beneficial than defect identification. (See also *Cost of Quality* and *Prevention*.)

For more information, see: Philip B. Crosby, *Quality Is Free* (New York: McGraw-Hill, 1979); Sarv Singh Soin, *Total Quality Control Essentials* (New York: McGraw-Hill, 1992); Armand V. Feigenbaum, *Total Quality Control,* Third Edition, Revised (New York: McGraw-Hill, 1991); V. Daniel Hunt, *Quality in America* (Burr Ridge, IL: Irwin Professional Publishing, 1992).

Roadblock Identification Analysis This technique is used to identify circumstances or other obstacles that do, or could, prevent improvements in performance or that block existing products or services from achieving specified levels of quality. Applying nominal group techniques, users identify issues and prioritize them by the degree to which they represent roadblocks to success. Then teams can perform analysis of key barriers and propose ways to remove them.

For more information, see: William Lareau, *American Samurai* (New York: Warner Books, 1991).

Robust Design This refers to a kind of design consciousness and approach that minimizes the possibility of failure and maximizes product reliability. It includes the use of parts and components that do not fail under extreme or noncontrollable circumstances. It incorporates approaches to assembly that make it difficult for a product to be improperly put together (poka-yoke). It emphasizes simple over complex design. Robust designs reduce the opportunity for defects to come from the production process and increase the probability that products will operate as they are supposed to for a long time in the face of many different types of uncontrollable circumstances.

> The principal idea in the Taguchi philosophy is that statistical testing of a product should be carried out *at the design stage* in order to make the product and the process *robust* to variations in the manufacturing and use environment.
>
> *N. Logothetis*

Product designers use statistical process control techniques to understand the effect of different variables on the overall performance of products with the intent of increasing reliability. In the early 1990s, process owners began to apply similar techniques to the design of processes to increase the probability of the process performing as required.

For more information, see: Joseph M. Juran and Frank M. Gryna, *Quality Planning and Analysis*, Third Edition (New York: McGraw-Hill, 1993); N. Logothetis, *Managing For Total Quality: From Deming to Taguchi and SPC* (Englewood Cliffs, NJ: Prentice Hall, 1992).

Robustness A recent entrant into the quality vocabulary, it refers to the condition or quality of a product, service, or process in which performance is relatively stable, with a minimum of variability, despite uncontrollable changes in the environment or operating conditions.

The term originated in product design engineering and then spread to information processing professionals designing large, complex software systems. Today the term is being used by process owners to mean the ability of a process to perform as required regardless of special causes or other environmental interferences.

For more information, see: Joseph M. Juran and Frank M. Gryna, *Quality Planning and Analysis*, Third Edition (New York: McGraw-Hill, 1993); Dev G. Raheja, *Assurance Technologies* (New York: McGraw-Hill, 1991).

Root Cause This refers to the underlying reason why a process or product does not conform to specifications or requirements. The identification and elimination of a root cause for a problem brings the process or product back to conformance. Identifying root causes early in a process helps to explain problems that happen much later in a process. This is an approach that emphasizes figuring out how to prevent problems as early as possible rather than dealing with problem symptoms later on. There are a number of techniques for finding root causes, including the cause-and-effect diagram, checksheets, and other SPC tools.

For more information, see: Paul F. Wilson, Larry D. Dell, and Gaylord F. Anderson, *Root Cause Analysis: A Tool for Total Quality Management* (Milwaukee, WI: ASQC Quality Press, 1993); H.G. Menon, *TQM in New Product Manufacturing* (New York: McGraw-Hill, 1992).

Root Cause Analysis An important quality tool for identifying the source of defects or problems, it is a structured approach that focuses on the ultimate or original cause of a problem or condition. The approach discounts snap judgments or dealing with problem symptoms and relies instead on a disciplined investigation of possible problem sources. It is one of the most widely used and effective approaches to process improvement.

Root cause analysis involves a variety of techniques—almost all simple to use—that define problems and lead to solutions. It also provides a basis for evaluating the effectiveness of those solutions. Techniques for investigating root causes include change analysis, barrier analysis, causal factors analysis, and cause-and-effect analysis, among others.

> For more information, see: Paul F. Wilson, Larry D. Dell, and Gaylord F. Anderson, *Root Cause Analysis: A Tool for Total Quality Management* (Milwaukee, WI: ASQC Quality Press, 1993); H.G. Menon, *TQM in New Product Manufacturing* (New York: McGraw-Hill, 1992).

> Structure in the problem-solving method provides for a systematic search for the root cause. It provides a system that is internally consistent and flexible, and also forms the basis for communication between the various groups involved in solving a particular problem.
>
> *H.G. Menon*

Rules of Conduct These are guidelines usually established by a team at the start of its process work to govern the behavior and performance of its members. Efficient teams always establish rules of conduct at the beginning of their work. Peter Scholtes, author of *The Team Handbook*, recommends the following guidelines for conducting team meetings to ensure that everyone participates and the team stays focused:

- *Ask for clarification in discussion.* This helps you make sure members don't misunderstand one another.
- *Act as gatekeepers.* This helps to throttle members who might tend to dominate the discussion.
- *Listen.* The only way you can learn what others have to say is to listen, with empathy.
- *Summarize.* This is another way to minimize misunderstanding.
- *Contain digression.* This keeps the group focused on the right issues.

- *Manage time.* By setting time limits for discussion and then keeping to these, you help ensure that the meeting will cover what it is supposed to in an efficient manner.
- *End the discussion.* By understanding when everyone has said what is on their mind, you can then summarize or make decisions based on discussion without dragging things out.
- *Test for consensus.* Summarize the group's position and then make sure everyone agrees.
- *Regularly evaluate the meeting process.* Look for opportunities for improvement.

For more information, see: Peter R. Scholtes, *The Team Handbook* (Madison, WI: Joiner Associates, 1988); Jack D. Orsburn et al., *Self-Directed Work Teams* (Burr Ridge, IL: Irwin Professional Publishing, 1990); Kimball Fisher et al., *Tips for Teams* (New York: McGraw-Hill, 1995).

Run Chart This is a graph that visually presents the plots of performance information over time for such things as processes or defects or any measured data collected through time.

Run charts are particularly useful in helping to monitor performance of a process over time to determine whether or not averages are changing. Points are plotted in the order in which they occur and lines connecting the dots from left to right are drawn. Run charts help detect any long-term trends or unusual patterns in process outputs through time. They help determine whether something might be wrong with a process, worthy of further study. They are the first step in preparing control charts. Figure 71 shows a typical run chart. In this case, it documents the average weights of coffee above or below the label weight on subsequent days in a particular month.

For more information, see: Michael Brassard and Diane Ritter, *The Memory Jogger II* (Methuen, MA: GOAL/QPC, 1994); N. Logothetis, *Managing For Total Quality: From Deming to Taguchi and SPC* (Englewood Cliffs, NJ: Prentice Hall, 1992); Hy Pitt, *SPC for the Rest of Us* (Reading, MA: Addison-Wesley, 1994).

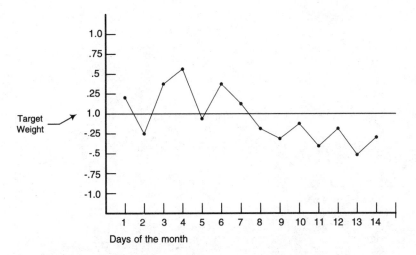

1.0
.75
.5
.25
1.0
-.25
-.5
-.75
-1.0

Target
Weight

1 2 3 4 5 6 7 8 9 10 11 12 13 14

Days of the month

FIGURE 71

A run chart showing
the weight in quarter
ounces above or below
the target weight on the
label for canned coffee
based on the average of
100 samples taken daily
over two weeks. Seven
points below the target
weight indicates a po-
tential problem with
the filling machine.

Running Control This is another way to describe process control as an ongoing activity at every step in the process using SPC tools and techniques. This emphasizes prevention and early detection of problems. On a flowchart, we may find several different control points for stop/continue decisions throughout the process for assuring that the output of that step meets requirements for the next step to occur. These points include actions to take when outputs do not meet requirements. Of course the goal of TQM and SPC is to minimize the possibility of defects emerging from any process step. Figure 72 (p. 322) shows an example of a running control point in a process.

> For more information, see: H.G. Menon, *TQM in New Product Manufacturing* (New York: McGraw-Hill, 1992); Eugene H. Melan, *Process Management* (New York: McGraw-Hill, 1993).

Runs These are patterns on a run chart (also called a control chart) in which a number of data points will line up on one side of a central line. Past a specified number of consecutive data points, a pattern will become unnatural and thus worthy of further study.

The most common runs are data on a line chart that has two axes. Figure 71 illustrates a simple run chart with the *x* axis indicating chronological time (in this case, consecutive days of the month) and the *y* axis indicating weights above or below a

FIGURE 72

A running control point as shown in a segment of a process flowchart.

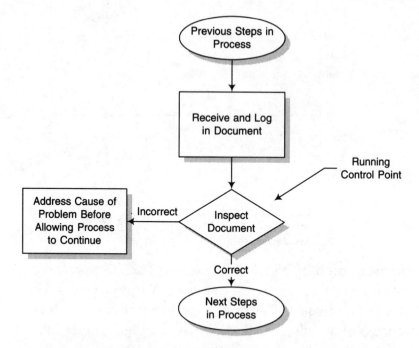

target. Each piece of data that contributed to the creation of this chart is part of a run.

For more information, see: N. Logothetis, *Managing For Total Quality: From Deming to Taguchi and SPC* (Englewood Cliffs, NJ: Prentice Hall, 1992); George L. Miller and LaRue L. Krumm, *The Whats, Whys & Hows of Quality Improvement* (Milwaukee, WI: ASQC Quality Press, 1992).

Sample This refers to a specific (finite) number of items of a similar type taken from a population or lot for the purpose of examination to determine if all members of the population or lot conform to quality requirements or specifications.

Sampling as a way of measuring has been an important method of testing quality for decades, primarily in the area of components, raw materials, and finished products. Sampling techniques are also proving effective in measuring performance of processes (for example, number of burned hamburgers, invoices with errors, unanswered telephone calls, and so on). The use of sampling techniques depends on statistical analysis, and the number of samples chosen depends on the sampling plan chosen. Statisticians have developed several such plans appropriate for different types of products.

While samples and sampling remain important in quality control, it is important to appreciate that most sampling procedures are set up to check outputs after the fact. They do not have anything to do with controlling processes so as to reduce or prevent defects in outputs. Statistical process control, conversely, is largely about using various tools and techniques for continuously reducing variation in outputs that result in defects in the first place. As such, SPC is an important part of an overall TQM effort in an organization. Such efforts also include always improving your ability to satisfy customers, using teams, developing a culture of cooperation, and many other changes that make a company more efficient and effective.

For more information, see: Joseph M. Juran and Frank M. Gryna, *Quality Planning and Analysis,* Third Edition (New York: McGraw-Hill, 1993); Armand V. Feigenbaum, *Total Quality Control,* Third Edition, Revised (New York: McGraw-Hill, 1991); Henry L. Lefevre, *Quality Service Pays: Six Keys to Success!* (Milwaukee, WI: ASQC Quality Press, 1989); George L. Miller and LaRue L. Krumm, *The Whats, Whys & Hows of Quality Improvement* (Milwaukee, WI: ASQC Quality Press, 1992); Hy Pitt, *SPC for the Rest of Us* (Reading, MA: Addison-Wesley, 1994).

Sample Selection This is the process used to select samples from a lot for testing of any type. The process typically is a random one intended to identify samples that would reflect the quality standards of the components, products, or activities being tested.

Some different methods for selecting samples include:

- *Cluster Sampling.* Divide the lot into clusters and choose one for testing.
- *Stratified Sampling.* Divide the lot into various proportions or strata and then select random samples from each strata. This may be used when lots are very large.
- *Selected Sampling.* This emphasizes sampling at certain times of the day or night. This is considered more precise than random sampling but may have some bias associated with it.

For more information, see: Hy Pitt, *SPC for the Rest of Us* (Reading, MA: Addison-Wesley, 1994); George L. Miller and LaRue L. Krumm, *The Whats, Whys & Hows of Quality Improvement* (Milwaukee, WI: ASQC Quality Press, 1992).

Sample Size This is a specific number of items randomly chosen from a lot or any population selected for testing. The sample size depends on the sampling plan chosen and the size of the lot being checked. The larger the sample, the more accurately it will represent the entire population. However, using statistics, we can determine a sample size above which any differences would be negligible. In general, the higher the quality standard, the larger the sample size would be. In other words, if your standard for acceptable is not more than .5 percent defects in any lot, the sample size would be larger than if your standard was 2.5 percent defects.

> For more information, see: Eugene L. Grant and Richard S. Leavenworth, *Statistical Quality Control*, Sixth Edition (New York: McGraw-Hill, 1988), Richard DeVor, Tsong-How Chang, and John W. Sutherland, *Statistical Quality Design and Control* (New York: Macmillan, 1992).

Sampling This is the process of collecting a sample of data about the quality of a lot or population of goods, actions, or information as reflective of the patterns of the entire population or lot. (See also *Sample*.)

> For more information, see: Hy Pitt, *SPC for the Rest of Us* (Reading, MA: Addison-Wesley, 1994); N. Logothetis, *Managing For Total Quality: From Deming to Taguchi and SPC* (Englewood Cliffs, NJ: Prentice Hall, 1992).

Sampling Fraction This is the ratio of a sample size to the total number of units in a population or lot. Units may be components, products, or services.

Sampling Frequency As the phrase suggests, it is the number of times samples are inspected during the course of a process. Historically, this kind of sampling is of products on a production line or components about to be used. In process quality control, it may also involve inspection of services. It is also the ratio of the number of units selected for inspection at a specific point in the production or service process to the number of units of product or services moving past that point of inspection.

> For more information, see: N. Logothetis, *Managing For Total Quality: From Deming to Taguchi and SPC* (Englewood Cliffs, NJ: Prentice Hall, 1992).

Sampling Interval This is any systematic sampling that occurs at specific points in time (for example, once every two hours) or after a specific quantity of output between samples (for example, sample one unit for every hundred that go through the line).

Scanlon Committee This refers to a committee made up of various levels of people in an organization who come together to implement a philosophy of management/labor relations with the intention of improving the quality of work life in the organization.

This committee is part of something known as the Scanlon Plan (named after Joseph Scanlon, a one time trade union official and later professor at MIT). This was one of the first attempts to bring employees and supervisory personnel together to work cooperatively on quality and to improve life on the job. His was the first widely recognized effort that proved workers are more committed to quality performance when they are directly involved in decisions involved in the performance of their work.

For more information, see: Armand V. Feigenbaum, *Total Quality Control*, Third Edition, Revised (New York: McGraw-Hill, 1991).

Scatter Diagram Also known as a cross plot, this is a widely used graphical tool. It is a technique employed to illustrate relationships between two variables. Two sets of data are plotted graphically, with the x axis representing a variable to make a prediction and the y axis to display the variable to be predicted. Figure 73 shows various patterns of data on scatter diagrams and what these patterns mean in terms of correlations between variables.

FIGURE 73

Scatter diagrams showing various patterns for data.

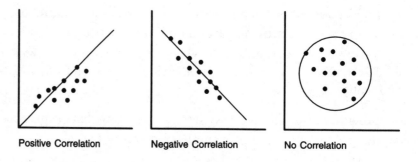

Positive Correlation   Negative Correlation   No Correlation

It is a useful tool for establishing if a relationship exists between two factors. Clusters of data create patterns. The tighter or more clustered that data is, the more probable the conclusion that the data is related. Another rule of thumb is that the straighter the line around which data clusters, the more likely that as one variable changes so too would the other variable by the same proportion. This tool is helpful in understanding possible cause-and-effect relationships and provides clues for how to improve quality by manipulating these factors to get the result you are looking for. However, it does not provide any conclusive sense of cause-and-effect, which must come from more thorough analysis. (See also *Cross Plot*).

> The causal relationship, if any exists at all between the variables, is not established by the scatter diagram; investigative work, including testing and experimentation, is required to determine if any causal relationship exists.
>
> *Hy Pitt*

> For more information, see: Michael Brassard and Diane Ritter, *The Memory Jogger II* (Methuen, MA: GOAL/QPC, 1994); James H. Saylor, *TQM Field Manual* (New York: McGraw-Hill, 1992); George L. Miller and LaRue L. Krumm, *The Whats, Whys & Hows of Quality Improvement* (Milwaukee, WI: ASQC Quality Press, 1992).

Selection Grid This is a tool used to make decisions by comparing alternative solutions or opportunities against each other. Another term for this is decision matrix. It lists solutions or problems in cells along the left vertical column and criteria for selection among them in cells along the top horizontal row. You then use a scoring system to select among alternatives those that score the highest. It is not meant to automate decision making but as a tool for better judging many alternative variables. See figure 30 on page 117 under the entry *Decision Matrix* for an example and more explanation.

> For more information, see: James H. Saylor, *TQM Field Manual* (New York: McGraw-Hill, 1992).

Self-Directed Teams Also known as self-managed teams, these are groups of employees who have taken ownership of a process or processes. They work with little direct management control. They have been given authority and responsibility to manage their part of the business, including, among other things, hiring

The self-managing team should become the basic organizational building block. Train them, recruit them on the basis of team-work potential; pay them for performance; and clean up bureaucracy around them, dramatically chang-ing the roles of middle managers and staff experts.

Tom Peters

authority, training new employees, disciplining team mem-bers, financial responsibility, scheduling, setting production tar-gets, undertaking improvement activities, and performing self-appraisals.

The rationale for moving to self-directed teams includes: (1) the recognition that work gets done as processes that require cooperation among employees; (2) the desire to take full advan-tage of the skills and talents of all employees; (3) the recognition that multiskilled employees working on teams can do several dif-ferent jobs making work go more smoothly; (4) the availability of technology that gives everyone access to information and the ability to communicate easily among themselves and others in the organization; and (5) this approach helps instill a sense of pride and commitment to their jobs and to the company, en-hancing their intrinsic motivation. With the availability of infor-mation and communication technology, this often obviates the need for a boss or middle manager to disperse and control infor-mation. Old rules of information transfer by managers often do not add value and create the superior-subordinate attitudes and role playing that often contribute to mediocrity and waste.

A company moving to self-managed teams must have a cul-ture that supports this approach. Companies must be willing to invest in training so team members have the knowledge and ex-perience that is commensurate with their responsibility. Such training includes TQM practices and tools, teamwork skills, meeting skills, communication skills, and hiring skills, and con-tinuously upgrading their job skills as well.

For more information, see: Charles C. Manz and Henry P. Sims, Jr., *Business Without Bosses* (New York: John Wiley & Sons, 1993); William C. Byham, *ZAPP: The Lightning of Empowerment* (New York: Fawcett Columbine, 1988); Tom Peters, *Thriving on Chaos* (New York: Alfred A. Knopf, 1987); Jack D. Orsburn, et al., *Self-Directed Work Teams: The New American Challenge* (Burr Ridge, IL: Irwin Professional Publishing, 1990).

Seven Tools of Quality These are the basic tools for process control and improvement. They include (1) the cause-and-effect diagram, (2) the checksheet, (3 the control chart, (4) the flowchart, (5) the histogram, (6) the Pareto chart, and (7) the scatter diagram.

The use of these tools is based on the understanding that (a) all work consists of tasks in processes, (b) there is always some variation in processes, and (c) improvement in the way processes operate and in the quality of outputs is always possible. These tools allow anyone to (1) document and measure what happens in a process so as to understand each step, (2) learn where complexities and redundancies lie, (3) distinguish between common cause and special cause variation, (4) identify the relationship between causes and effects, and (5) identify leverage points where the greatest improvements will come from one's efforts, and (6) make continuous improvements that reduce variation and prevent problems from occurring in the first place.

> For more information, see: Hy Pitt, *SPC for the Rest of Us* (Reading, MA: Addison-Wesley, 1994); George L. Miller and LaRue L. Krumm, *The Whats, Whys & Hows of Quality Improvement* (Milwaukee, WI: ASQC Quality Press, 1992); Michael Brassard and Diane Ritter, *The Memory Jogger II* (Methuen, MA: GOAL/QPC, 1994); PQ Systems, Inc., *Total Quality Transformation Improvement Tools* (Miamisburg, OH: Productivity Quality Systems, Inc., 1994; 800-777-3020); Nancy R. Tague, *The Quality Toolbox* (Milwaukee, WI: ASQC Quality Press, 1995).

Sharing Rallies These are meetings in which individuals, but usually teams, are recognized for their contributions and successes with processes. Often these are recognition events with awards, motivational speakers, and trips and are celebrations of progress. Those honored typically explain to their audiences what they did and what they learned from the effort.

> For more information, see: Jack D. Orsburn, et al., *Self-Directed Work Teams: The New American Challenge* (Burr Ridge, IL: Irwin Professional Publishing, 1990).

Shewhart Cycle This is another name for plan-do-check-act continuous improvement cycle. W. Edwards Deming always referred to it in this way as he attributed its development to Walter Shewhart, who also developed the statistical control chart. (See also *Plan-Do-Check-Act* and *Continuous Improvement*.)

> For more information, see: W. Edwards Deming, *Out of the Crisis* (Cambridge, MA: MIT Center for Advanced Engineering Study, 1986).

Sigma (σ) This is the accepted mathematical symbol for standard deviation in a distribution of values in population. (See *Standard Deviation* for more.)

Signal-to-Noise Ratio (S/N Ratio) Borrowed from communication theory, this is the ratio of the mean (the signal) in a process to the standard deviation (noise). In other words, it is a way of quantifying how well the process is operating.

What it captures is how much variation there is in a process. A lower S/N ratio indicates there is much variation in a process and a higher S/N ratio indicates less variation. In general we want to reduce variation and want a high S/N ratio. However, this may not always be the case, and many processes have uncontrollable factors that make a low S/N ratio not possible. This understanding has become important in the development of products and processes, allowing engineers to develop more robust designs that withstand uncontrollable factors. Despite variation, robust processes deliver outputs that meet requirements. Robustly designed products also operate properly in a wide variety of conditions.

> For more information, see: George L. Miller and LaRue L. Krumm, *The Whats, Whys & Hows of Quality Improvement* (Milwaukee, WI: ASQC Quality Press, 1992); Phillip J. Ross, *Taguchi Techniques for Quality Engineering* (New York: McGraw-Hill, 1988).

Six Sigma This is one of the most popular forms of measuring performance in manufacturing and is also widely used in large corporations. Six Sigma means 3.4 defects, errors or failures per million occurrences or performances.

Motorola, an early Baldrige Award winner, popularized this measure of quality. It has to do with the capability of a process.

The higher the process capability index, the better. A process capability index of 2 indicates that there will be zero outputs that fall outside specification limits, a very desirable situation. The idea of Six Sigma comes from the assumption that whatever your specification limits are, this will be set at Six Sigma above and below the process mean. With those as specification limits, you then devise a process such that even if the process average shifts 1.5 sigma (standard deviations) to the right or left of the process mean, these shifts will still yield only 3.4 defects per million. Figure 74 illustrates this concept.

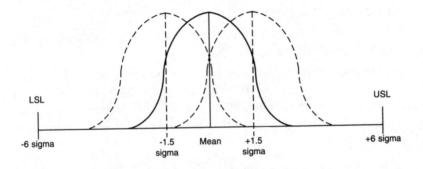

FIGURE 74

Six Sigma assumes that specification limits are plus or minus Six Sigma distance from the process average, and even if the process average shifts 1.5 sigmas right or left, the process will still deliver only 3.4 defects per million.

Implementing Six Sigma quality requires not only careful measure and management of processes, but also products with robust design that operate well with parts that fall within various tolerance ranges. In other words, defect-free products depend both on design and on well-managed processes. Poka-yoke and robust design are two ways that companies can approach Six Sigma quality. (See *Poka-Yoke* and *Robust Design*.)

For more information, see: Joseph M. Juran and Frank M. Gryna, *Quality Planning and Analysis*, Third Edition (New York: Mc-Graw-Hill, 1993).

Skills Management This refers to the management of employee skills in terms of experiences, competencies, values, attitudes, and patterns of behavior. Often formally catalogued, skills provide an organization with an inventory of capabilities and people, and a list of possible areas for further training and practice.

As companies move more toward the implementation of Total Quality Management and teams, it becomes important to continuously upgrade technical, administrative, and interpersonal skills. This allows a company to maximize the productivity of its employees and to stay competitive in an ever-changing marketplace. W. Edwards Deming recognized the importance of companies paying attention to the skills of their employees and made that the focus of two of his 14 Points for Management: Point 6, "Institute training on the job," and Point 13. "Institute a vigorous program of education and self-improvement." In other words, the notion of continuous improvement applies not just to process and products but also to the skills of those responsible for managing those processes and delivering those products.

> For more information, see: Jack D. Orsburn, et al., *Self-Directed Work Teams: The New American Challenge* (Burr Ridge, IL: Irwin Professional Publishing, 1990); W. Edwards Deming, *Out of the Crisis* (Cambridge, MA: MIT Center for Advanced Engineering Study, 1986).

Special Cause This is a source of variation in the performance or output of a process that is unpredictable and not due to causes as the system normally operates. Engineers will often also use the phrase *assignable cause* to mean the same thing, implying the cause of the variation is an exception.

> A control chart detects the presence of a special cause but does not *find* the cause—that task must be handled by a subsequent investigation of the process.
>
> *Joseph M. Juran and Frank M. Gryna*

The idea of special cause variation is very important in statistical process control and the use of control charts. A special cause is indicated on a control chart by a measurement that falls outside the control limits. For example, if a manufacturing process delivers an average of 100 units per hour and then dips to 75 for one hour, you can expect that is because of a special cause, such as a power outage or an unanticipated shortage of parts or some similar special and identifiable cause. Such problems are often easy to solve (or at least explain). They are usually one-time happenings. Addressing special cause problems does not do anything to improve variation due to common causes that have to do with the interactions of people, machines, and the operation of the system as a whole.

A major problem in traditional management is that managers do not track processes and treat any variation as if it were due to a special cause. For example, if the average output of a process is 100 units per hour one week and 97 units per hour the next week, this is likely just due to variation within the process itself, which would be easily documented using a control chart. However, a traditional manager may get upset by this drop, treat it as a special cause problem (when it is not), and start making changes that introduce new elements to the system that can cause even greater variation. Deming called this *tampering* (see this term) and railed against it in his books and seminars.

> For more information, see: W. Edwards Deming, *Out of the Crisis* (Cambridge, MA: MIT Center for Advanced Engineering Study, 1986); Brian L. Joiner, *Fourth Generation Management* (New York: McGraw-Hill, 1994); Joseph M. Juran and Frank M. Gryna, *Quality Planning and Analysis*, Third Edition (New York: McGraw-Hill, 1993).

Specification This is the engineering requirement for judging the acceptability of a product or some aspect of product in terms of appearance, durability, performance, size, and so on.

This is an important concept in quality management. Meeting customer requirements depends on delivering outputs that adhere to specifications. One of the reasons Japanese firms have excelled in manufacturing high-quality products is their approach to meeting specifications. In the past in the United States, this has meant delivering parts, for example, that fall within some defined tolerances for variation from piece to piece. As long as an item falls within those tolerances, it was considered to meet specifications and the process to be operating well. This approach works fairly well. It is known as the "goal post" approach from football. The Japanese approach is to develop a stated target specification, and then to consider any output that does not meet this specification to be a possible loss for the customer and for the company itself. To achieve this target specification, they continuously improve their processes to remove as much variation as possible. (See *Loss Function* for more on this idea.)

> For more information, see: W. Edwards Deming, *Out of the Crisis* (Cambridge, MA: MIT Center for Advanced Engineering Study, 1986); Brian L. Joiner, *Fourth Generation Management* (New York: McGraw-Hill, 1994); Phillip J. Ross, *Taguchi Techniques for Quality Engineering* (New York: McGraw-Hill, 1988).

Specification Limits These are the measurements within which an output must fall or the requirements a product or component must meet to be considered acceptable. It is important not to confuse specification limits with control limits. An in-control process may or may not be able to deliver outputs that meet specifications. Using statistical process control charts, we can calculate the ability of a process to deliver outputs that meet specifications. This is the capability process index. (See *Capability Process Index* for more on this.) Figure 75 shows the difference between a process that exceeds specification limits and one that does not.

FIGURE 75

This shows the distribution of process outputs for a process that can and a process that cannot operate within specification limits.

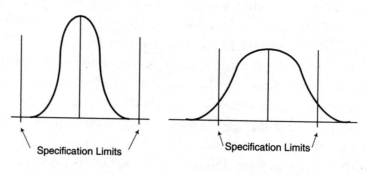

A Highly capable process where all output fall substantially within specification limits

A process in which a part of the output will always fall outside specification limits.

Quality experts remind us that just making sure the process can deliver within specification limits is not enough. It is important to continue to refine and improve processes, which reduces sources of variation, lowers costs, and increases quality and customer satisfaction.

> For more information, see: W. Edwards Deming, *Out of the Crisis* (Cambridge, MA: MIT Center for Advanced Engineering Study, 1986); Brian L. Joiner, *Fourth Generation Management* (New York: McGraw-Hill, 1994); George L. Miller and LaRue L. Krumm, *The Whats, Whys & Hows of Quality Improvement* (Milwaukee, WI: ASQC Quality Press, 1992).

Stakeholders These include individuals and organizations that have an investment or interest (that is, a stake) in the success or actions taken by an organization. Stakeholders include customers, managers and employees, competitors, unions, stockholders, business partners, suppliers, the community, the nation, the media, and so on.

> Every organization has major constituencies or stakeholders whose needs are ignored at the organization's peril.
>
> *Burt Nanus*

The idea of stakeholders has taken on greater prominence in those organizations that are implementing quality management. One reason is because quality management looks on the world from the systems view, which emphasizes interrelationships and interdependencies. One important task of management is to understand these relationships and dependencies and balance them to the benefit of all. This balancing process suggests that you cannot focus just on satisfying customers or just on maximizing short-term profit for stockholders or just on the creation of a positive workplace for employees. However, TQM, which emphasizes teamwork and efficiency in the workplace and increasing the value offered to customers, can bring about greater satisfaction for all stakeholders than traditional management can. This is because traditional approaches tend to push short-term profitability, often at the expense of customers, employees, the community, and other stakeholders.

> For more information, see: Burt Nanus, *Visionary Leadership* (San Francisco: Jossey-Bass Publishers, 1992).

Standard This is a statement, specification, or quantity of material against which measured outputs from a process may be judged as acceptable or nonacceptable.

In a *normal curve*, approximately 68.2 percent of all data lie within 1 standard deviation of the mean; approximately 95.5 percent will be within 2 standard deviations, and 99.7 percent will be within 3 standard deviations.

George L. Miller and LaRue L. Krumm

Standard Deviation This is the average difference between any measured data point in a population and the mean of all those points taken together. In relative terms, the closer the standard deviation is to the mean, the more likely it is that any single point in the population is around the same value as the mean itself. Said another way, the higher the standard deviation (usually symbolized by the Greek letter sigma, σ), the more variation there is in the process. The lower the standard deviation, the less variation there is in the process. Standard deviation, then, is a way a learning how well the process is operating. (Calculating the standard deviation is straightforward, following an arithmetic formula, found in all books on this subject. More important, however, is an understanding of its meaning in statistical process control.) A goal for any process is to continually reduce variation in outputs, with the understanding that you can never get rid of it entirely.

The standard deviation is important in developing the bell curve for the distribution of measurements in any process. Statisticians discovered that in any population, they can create a curve, under which 99.7 percent of all measurements will fall within 3 standard deviations above or below the mean for the population. This same insight led to the development of control charts with upper and lower control limits placed 3 standard deviations above and below the center line (or mean). A bell curve (a standard distribution) that is high and narrow indicates that there is little variation in the process. A bell cure that is low and wide indicates that a process has a lot of variation. Figure 76 illustrates these differences.

For more information, see: Hy Pitt, *SPC for the Rest of Us* (Reading, MA: Addison-Wesley, 1994); George L. Miller and LaRue L. Krumm, *The Whats, Whys & Hows of Quality Control* (Milwaukee, WI: ASQC Quality Press, 1992); Armand V. Feigenbaum, *Total Quality Control*, Third Edition, Revised (New York: McGraw-Hill, 1991).

Statistic This is a quantity arrived at by taking samples from a population to arrive at some conclusion about the overall characteristics of that population.

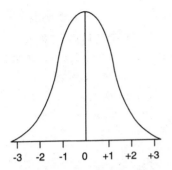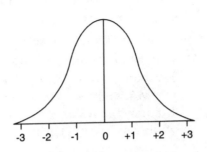

FIGURE 76

Two normal distributions showing the difference between lower and higher standard deviations. Each curve extends 3 standard deviations to the right and left of the mean.

A population distribution in which standard deviation is small, resulting in a tall narrow bell curve

A population distribution in which standard deviation is larger, resulting in a short wide bell curve

Statistical Control This describes the status of a process from which all special causes of variation have been eliminated, leaving only common causes.

When a process is said to be under statistical control, that is another way of suggesting it is stable. This does not suggest that the outputs of that process will meet specifications or customer requirements. It is the responsibility of management and everyone in the organization to then address the process as a whole, the interdependencies among people, machines, and materials, to learn how to reduce common cause variation, which improves quality and lowers costs.

For more information, see: W. Edwards Deming, *Out of the Crisis* (Cambridge, MA: MIT Center for Advanced Engineering Study, 1986); Brian L. Joiner, *Fourth Generation Management* (New York: McGraw-Hill, 1994); Hy Pitt, *SPC for the Rest of Us* (Reading, MA: Addison-Wesley, 1994).

Statistical Methods This refers to the application of the theory of probability of variation in any system. Basic statistical methods are tools and techniques widely used in the management of processes. These methods use such popular items as control charts, process capability analysis, and statistical inference. A more sophisticated set of tools, less widely used in normal process work, involves statistical analysis and design of experiments, regression analysis, correlation analysis, and variance analysis. Both sets of tools are more commonly evident today in all types of processes and organizations as managers come to appreciate that all work involves processes, and these tools are how they understand and improve how processes operate.

For more information, see: Hy Pitt, *SPC for the Rest of Us* (Reading, MA: Addison-Wesley, 1994); George L. Miller and LaRue L. Krumm, *The Whats, Whys & Hows of Quality Improvement* (Milwaukee, WI: ASQC Quality Press, 1992).

Statistical Process Control (SPC) This is the body of statistical techniques used to measure and monitor the performance of processes. The reason for applying these is to identify specifically areas for improvement in processes and to measure variation in outputs of processes, all leading to actions that will reduce variation in outputs. The objectives of SPC include (1) making processes stable (that is, subject only to common cause variation) and (2) continuously improving their capability to deliver outputs that meet a target specification or requirement with minimal variation from this target.

> It's always very enlightening to demonstrate to management (the ones who are interested—don't get pushy) that what looks like significant changes in a process are only normal variation.
>
> *William Lareau*

Another, more general, way of describing SPC is to see it as a management philosophy that is a disciplined approach to the identification and resolution of process problems. Popular SPC tools include fishbone diagrams, graphical representations, Pareto charts, and control charts. All are used to track numerically what a process is doing, analyzing that performance, and suggesting improvements.

Many people have come to see Total Quality Management mainly as the application of SPC tools. However, this represents a deep misunderstanding of TQM. TQM involves understanding organizations as systems with processes for serving customers. The use of SPC and its tools is simply the logical outcome of this insight into organizations and their purpose.

For more information, see: W. Edwards Deming, *Out of the Crisis* (Cambridge, MA: MIT Center for Advanced Engineering Study, 1986); Hy Pitt, *SPC for the Rest of Us* (Reading, MA: Addison-Wesley, 1994); William Lareau, *American Samurai* (New York:

Warner Books, 1991); Brian L. Joiner, *Fourth Generation Management* (New York: McGraw-Hill, 1994).

Statistical Quality Control (SQC) This is a broader term than SPC, implying the use of statistical techniques to measure and improve processes and quality. While SPC is mainly concerned with controlling processes to reduce variation and improve quality, SQC incorporates the entire body of statistical practices for improving output quality, including SPC, inspection and sampling plans, and other statistical tools. TQM emphasizes SPC and prevention rather than inspection and sampling, which is an after-the-fact approach to quality control and improvement.

For more information, see: George L. Miller and LaRue L. Krumm, *The Whats, Whys & Hows of Quality Improvement* (Milwaukee, WI: ASQC Quality Press, 1992); Eugene L. Grant and Richard S. Leavenworth, *Statistical Quality Control*, Sixth Edition (New York: McGraw-Hill, 1988).

Statistics This is a branch of applied mathematics. Its purpose is to scientifically describe and analyze observations so as to reduce uncertainty about our ability to predict future events. This analysis improves the soundness of our decisions.

The techniques of statistics rely on the theory of probability, which makes it possible to study problems mathematically. Statistics was first introduced into manufacturing processes in the 1920s and after World War II began to appear in office processes. Today the use of statistics is an integral part of any Total Quality Management effort.

Those involved in quality management use two types of statistics. First, there is descriptive statistics, which we use to summarize, simplify, and present large masses of data. This branch uses tables of percentages, bar charts, pie charts, and so on to represent this data. This type of statistics involves breaking data into logical groups that have meaning for making decisions. Second, there is inferential statistics, which we use to make decisions about entire populations based on information derived from statistical analysis of limited samples from that population. Much of SPC is based on the application of inferential statistics.

> For more information, see: Robert Rosenfeld, *The McGraw-Hill 36-Hour Business Statistics Course* (New York: McGraw-Hill, 1992); or any of several books that deal with basic statistics.

Steering Committee or Group This is a committee, usually made up of senior management, that oversees and guides a cross-functional or focus activity. It is a widely used mechanism for coordinating the quality improvement activities of process and project teams across the entire organization. The job of this group or team is to provide direction for teams across the organization in terms of their missions, to get teams properly launched, to clear roadblocks to successful performance, and to provide resources needed for successful outcomes. Figure 77 shows the relationship between a steering committee and process and project teams.

FIGURE 77

The relationship between a steering committee and process teams. The two-headed arrows represent not a hierarchical but a cooperative relationship.

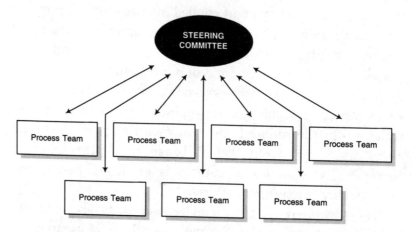

> For more information, see: Peter R. Scholtes, *The Team Handbook* (Madison, WI: Joiner Associates, 1988); Charles C. Manz and Henry P. Sims, Jr., *Business Without Bosses* (New York: John Wiley & Sons, 1993); James H. Saylor, *TQM Field Manual* (New York: McGraw-Hill, 1992).

Strategic Quality Management This approach to management goes beyond a concern with quality control and assurance. It is based on an understanding that there is a clear link between profitability and the continuous improvement of quality to maintain competitive advantage. In other words, it is a set of assumptions about what any organization must do to grow and prosper. It is not a defensive stance (like quality assurance, for example) but a comprehensive and proactive approach for helping any company survive and thrive in a very competitive environment. Said another way, it is a fundamental understanding of the strategic potential of TQM.

For more information, see: David A. Garvin, *Managing Quality* (New York: The Free Press, 1988).

Strategy This refers to the broad set of actions an organization decides on that will allow it to profitably use its resources to serve some group of customers. The best strategies involve a realistic assessment of organizational resources and capabilities, customer needs and expectations, and the environment. Total Quality Management provides the soundest foundation for making strategic decisions because it provides such realistic direction for what a company should be doing. This includes the continuous collection of information about customers and their needs and the continuous improvement of company resources to profitably deliver ever more value to customers.

For more information, see: W. Edwards Deming, *Out of the Crisis* (Cambridge, MA: MIT Center for Advanced Engineering Study, 1986); Brian L. Joiner, *Fourth Generation Management* (New York: McGraw-Hill); David A. Garvin, *Managing Quality* (New York: The Free Press, 1988).

> For the first time, top managers, at the level of presidents and chief executive officers, have expressed an interest in quality. They have linked it with profitability, defined it from the customer's point of view, and required its inclusion in the strategic management process. In the most radical departure of all, many have insisted that quality be viewed as an aggressive competitive weapon.
>
> *David A. Garvin*

Sub-Optimization In the application of quality management, this refers to optimizing the individual parts or functional areas of the organization at the expense of optimizing the whole organization. In other words, the irony of optimizing the parts means that you sub-optimize the entire organization. Since every functional area ultimately contributes to the delivery of quality to customers, only by cooperating and optimizing relationships among functions can a company avoid sub-optimization of its

In market-based societies, the dominant tradition has been for each company to look out for its own interests. That same urge is also present within companies. It is fostered by the prevailing systems of departmental goals and by the practice of conferring departmental monopolies with respect to decision making.

Joseph M. Juran

operations. In a dynamic, changing environment, no company can completely optimize its operations. There will be opportunities for improvement and "further optimization."

W. Edwards Deming recognized the tendency of organizations to departmentalize work according to function and the inefficiencies and sub-optimization this introduces into the organization as a whole. Point 9 of his 14 Points addresses this: "Break down barriers between staff areas." Only by breaking and eliminating such barriers and developing cross-functional co-operation can a company get at the waste that comes from departments operating as independent entities. A good part of process reengineering is about understanding how to do this by carefully examining how things are now and then replacing current processes. These bring all the people and functions together in ways that are simpler, more efficient, and better serve customers. (See also *Process Reengineering* and *Reengineering*.)

> For more information, see: W. Edwards Deming, *Out of the Crisis* (Cambridge, MA: MIT Center for Advanced Engineering Study, 1986); Michael Hammer and James Champy, *Reengineering the Corporation* (New York: HarperBusiness, 1993); Joseph M. Juran, *Juran on Leadership for Quality* (New York: The Free Press, 1989); William Lareau, *American Samurai* (New York: Warner Books, 1991).

Subprocess This is a part of a larger process that can be seen as almost a small process in itself. For example, in a process for renewing a driver's license, a subprocess might be those activities surrounding the written examination or the eye test. (See also *Macro-Process*.)

> For more information, see: H. James Harrington, *Business Process Improvement* (New York: McGraw-Hill, 1991).

Supplier This is any person or organization who provides input into a process. This can be a high school supplying students to a college, a company delivering raw materials to a factory, or another department providing you with data that you need to do your work. Suppliers are measured on effectiveness and are brought into your processes as participants rather than simply as vendors. They may be others within your own enterprise or vendors from other firms.

A supplier always has a customer. If you are a supplier, it is part of your responsibility to make sure you understand how you can deliver on the four utilities (form/function, place, time, and possession) that will best satisfy your customers. That helps develop the long-term cooperation to improve process operation and output quality. If you are a customer, it is your responsibility to view your suppliers (internal and external) as partners in your quest to deliver continuously improving outputs to your own internal customers and to the organization's final consumers. (See also *Customer Delight* and *Customer Requirements.*)

> For more information, see: Richard J. Schonberger, *Building a Chain of Customers* (New York: The Free Press, 1990); Michael Hammer and James Champy, *Reengineering the Corporation* (New York: HarperBusiness, 1993).

Supplier/Customer Analysis This technique is used to gather and exchange data on the needs and requirements of an organization to its key suppliers. It emphasizes defining the relationship between customer needs and supplier capabilities. The goal is to understand how customers and suppliers can work best together to their mutual benefit. (See also *Customer/Supplier Analysis.*)

> For more information, see: James H. Saylor, *TQM Field Manual* (New York: McGraw-Hill, 1992).

Supplier Quality Assurance A very obvious concept today, it is the expression of confidence that a supplier's quality will match a customer's requirements. This is brought about by a relationship between supplier and customer defined by processes that ensure a standard, routine high level of quality of supplies that meet the customer's needs. This allows a manufacturer, for example, to work with fewer suppliers. Joseph M. Juran defines nine essential activities to ensure supplier quality assurance:

> In the past, suppliers and buyers were often adversaries; some purchasers viewed suppliers as potential criminals who might try to sneak some defective products past the purchaser's incoming inspection. Today, the key word is partnership, i.e., working closely together for the mutual benefit of both parties.
>
> *Joseph M. Juran and Frank M. Gryna*

1. Define product and service quality requirements.
2. Evaluate alternative suppliers.
3. Select suppliers.
4. Conduct joint quality planning exercises.
5. Cooperate with supplier during the execution of a contract.
6. Gather proof of conformance to requirements.
7. Certify approved (qualified) suppliers.
8. Implement quality improvement programs as needed.
9. Create and use supplier quality ratings.

For more information, see: Joseph M. Juran and Frank M. Gryna, *Quality Planning and Analysis*, Third Edition (New York: Mc-Graw-Hill, 1993).

System This is a holistic view of an organization as a set of interrelated and interdependent parts, each of which, by its behavior, affects the behavior and performance of the others. In managing an organization as a system, you stop looking at what one part or the other is doing and concentrate on the *patterns of interaction* and how these patterns affect each other and the performance of the whole company. You make improvements in the system by modifying these patterns of interaction in ways that reduce redundancies and complexity, improve efficiency, and modify relationships that get in the way of high performance.

What is important in the system view is the *organization* of pieces and not the pieces themselves. This view helps us understand that success of any one part depends on what lots of other parts do. You appreciate that work does not get done by individuals working as independent agents but by processes that individuals, working together, execute. All this contrasts with the idea that an organization is a collection of independent entities that must somehow work together to achieve goals. This perspective causes you to focus on individuals rather than relationships and processes.

Understanding an organization as a system is at the heart of the successful implementation of Total Quality Management. It

> Systems thinking is a discipline for seeing wholes. It is a framework for seeing interrelationships rather than things, for seeing patterns of change rather than static "snapshots." It is a set of general principles—distilled over the course of the twentieth century, spanning fields as diverse as the physical and social sciences, engineering, and management.
>
> *Peter M. Senge*

gives meaning to an emphasis on processes (the steps by which inputs are transformed into outputs in a system), teamwork, continuous improvement, and customer focus. As soon as you understand an organization as a system, then the ideas of TQM become logical.

W. Edwards Deming based his view of management on what he called his theory of "Profound Knowledge." One of the cornerstones of this theory is the idea of organizations as systems. Two of the requirements for a system, as he explained it, include the idea that a system must have an aim or purpose, and that the system must be managed. The secret to that management, he said, is "cooperation between components toward the aim of the organization" (p. 54, *The New Economics*). Managing a system successfully means bringing together inputs from outside suppliers, transforming and adding value to those by various processes, and delivering outputs to customers who value them. Doing this well brings in the profit necessary to continue in business. All of Deming's work is fundamentally about how you use various tools and techniques to manage a system.

Peter Senge, author of *The Fifth Discipline: The Art and Science of the Learning Organization*, puts the system view at the heart of his teaching. Senge and others have studied systems and developed a series of "archetypes" that characterize different patterns of behavior in organizations. Using these archetypes, a manager can more effectively study organization problems and figure out what changes will have the greatest leverage for solving the problem.

For more information, see: W. Edwards Deming, *The New Economics* (Cambridge, MA: MIT Center for Advanced Engineering Study, 1993); Peter M. Senge, *The Fifth Discipline* (New York: Doubleday Currency, 1990); Peter M. Senge et al., *The Fifth Discipline Fieldbook* (New York: Doubleday Currency, 1994); Draper L. Kaufmann, Jr., *Systems One: An Introduction to Systems Thinking* (Minneapolis, MN: S.A. Carlton, Publisher, 1980, 612-920-0060); Daniel H. Kim, *Systems Archetypes* (Cambridge, MA: Pegasus Communications, 1992; 617-576-1231).

Taguchi Methods Named after Genichi Taguchi, a leading Japanese expert on quality improvement, these methods comprise a variety of techniques for evaluating quality and figuring out how to improve it.

Taguchi based his methods on the idea that any variation from customer requirements represents a loss to customers and to the company. With that as his starting point, he developed a variety of methods for managing and improving processes that minimize loss to all parties. These methods revolve around product and process design and operation. Taguchi methods optimize the relationships between customers, designers, and producers to the benefit of all. (See also *Loss Function* and *Quality Loss Function.*)

> For more information, see: Phillip J. Ross, *Taguchi Techniques for Quality Engineering* (New York: McGraw-Hill, 1988); H.G. Menon, *TQM in New Product Manufacturing* (New York: McGraw-Hill, 1992).

Tampering This the act of reacting to a common cause variation in a process as if it were a special cause variation. In other words, tampering is taking some action to adjust the system in response to a single data point you think happens because of a special cause. However, that point really represents one output of an in-control system.

To best understand tampering, first consider what it is not: If a worker continually makes a mistake on one kind of weld in a manufacturing process, it should be clear that this is due to an identifiable cause, for example, improper training. Retraining the worker should eliminate this particular variation or problem in output. This is a special cause problem, and fixing it in this way is appropriate and is not tampering.

Consider managers who review the output of an auto assembly process and discover that on one process there was an average of three scratches on a car door coming off the line. Using this single piece of data, they order that workers change their methods in some way these managers think (but have no real evidence) will eliminate this problem. In doing this, they are not aware that three scratches on a door falls within the control limits of how this assembly process normally operates. By modifying worker methods in this reactionary manner, they cannot predict what other variations might be introduced by the new methods, which affect the entire process. This is tampering. It also often happens when managers see some variance from budget, for example, and order a search for a cause, never appreciating that the variance is within expected control limits for how the system operates. Such efforts are wasteful, and we can never find a single cause for such variances; they are inherent in the system.

Tampering is a fundamental sin of traditional management. Deming often spoke out against tampering. He pointed out that it is making decisions based on hunches, guesswork, and single events rather than than a thorough understanding of process operation and capability. You can *never* improve a process by tampering.

> For more information, see: W. Edwards Deming, *The New Economics* (Cambridge, MA: MIT Center for Advanced Engineering Study, 1993); Brian L. Joiner, *Fourth Generation Management* (New York: McGraw-Hill, 1994).

> Of all the sources of tampering, the most deadly is management asking for explanations when in fact the variation is due to common causes.
>
> *Brian L. Joiner*

Task In understanding processes and their components, a task is a single action or substep in a process. In the analysis of processes, a task is generally the smallest identifiable action.

Team Much as in sports, this is a group of employees who work together to achieve a common objective. They are typically working on a process or project either in a single department or across functions. Team members rely on personal skills applied in a collaborative environment that fosters teamwork and consensus decision-making practices.

Teams and teamwork are an important part of the quality approach to management. Teams facilitate cooperation and the coordination of the different subprocesses and tasks by which final outputs are delivered to customers. They aid in communication and help minimize duplication of effort. They help eliminate destructive internal competition and contribute to a culture where everyone focuses on working together to the company's and customers' benefit.

Though the use of teams has many positive benefits, working in teams requires training in a variety of skills, including communication, meetings, group decision making, the use of improvement tools, and how to handle difficult people. In putting teams together, they often go through four stages: (1) forming, (2) storming, (3) norming, and (4) performing. Here is a brief description of each of these stages.

1. *Forming.* Here members explore what is acceptable group behavior. This is a transition from individual to team status.

2. *Storming.* Here the participants metaphorically jump in the water and check out their ability to swim. They are eager for progress but still inexperienced and not sure what actions to take. It can be the most tentative of the stages.

3. *Norming.* Here members start to get used to one another and develop some ground rules and their roles on the

> Rarely does a single person have enough knowledge or experience to understand everything that goes on in a process. Therefore, major gains in quality and productivity most often result from teams—a group of people pooling their skills, talents, and knowledge.
>
> *Peter R. Scholtes*

team. They start to feel they can get things done and start to settle down to work together.

4. *Performing.* At this stage the team is working together and starting to get results. They are learning from one another what works and feel an identification with each other as teammates.

For more information, see: Peter R. Scholtes, *The Team Handbook* (Madison, WI: Joiner Associates, 1988); Steve Levit, *Quality Is Just the Beginning* (New York: McGraw-Hill, 1994); Charles C. Manz and Henry P. Sims, Jr., *Business Without Bosses* (New York: John Wiley & Sons, 1993); Jack D. Orsburn et al., *Self-Directed Work Teams* (Burr Ridge, IL: Irwin Professional Publishing, 1990).

Tolerance Design This is a stage in any design process that defines the tolerances of variation in performance that will be acceptable in a final product or service. At this critical phase, where costs of end products can be increased dramatically, designers focus on producing robust designs and products in earlier parameter design phases to contain possible expenses. (See also *Robust Design* and *Poka-Yoke*.)

For more information, see: Joseph M. Juran and Frank M. Gryna, *Quality Planning and Analysis*, Third Edition (New York: McGraw-Hill, 1993).

Top-Down Flowchart This kind of flowchart lists the major steps in a process across the top of the page and then under each step lists its tasks. It is mainly used to describe the big picture of a process and get all the tasks down in a single place. It does not show complexity and redundancy in the process nor decision points or the processing of documents. It is easy to draw because it consists only of boxes with lists of tasks. It can also be helpful in thinking about new processes and the tasks they may entail. Figure 78 shows a top down flowchart for the development of this book.

For more information, see: PQ Systems, Inc., *Total Quality Transformation Improvement Tools* (Miamisburg, OH: Productivity

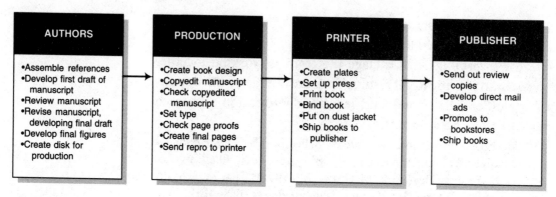

AUTHORS	PRODUCTION	PRINTER	PUBLISHER
•Assemble references •Develop first draft of manuscript •Review manuscript •Revise manuscript, developing final draft •Develop final figures •Create disk for production	•Create book design •Copyedit manuscript •Check copyedited manuscript •Set type •Check page proofs •Create final pages •Send repro to printer	•Create plates •Set up press •Print book •Bind book •Put on dust jacket •Ship books to publisher	•Send out review copies •Develop direct mail ads •Promote to bookstores •Ship books

FIGURE 78

A top-down flowchart for the development and publication of a book.

Value-driven leadership is essential to developing TQM and must be a top priority for upper management. It cannot be delegated or ignored.

Thomas H. Berry

Quality Systems, Inc., 1994; 800-777-3020); Peter R. Scholtes, *The Team Handbook* (Madison, WI: Joiner Associates, 1988).

Top Management Commitment This refers to the commitment and personal involvement of senior management in the implementation of quality programs. Their roles can include personal participation in quality committees, creation of quality processes, goals, measures, and policies, providing training on quality, and allocation of appropriate resources. Other activities include participation in process teams, inspection of quality progress, rewarding those who have contributed to quality improvements and who are "champions" of quality.

The idea of top management commitment is vital to the success of any quality management initiative. This is because top managers create the environment to which everyone else in the company adapts. If they suggest that TQM is a good idea and then assign its implementation to middle managers while not changing their own values and behavior, TQM will not happen in those companies. The value being communicated by this approach is "pay lip service to TQM, but keep doing what you were doing before." Why? Because this is what top managers are doing.

TQM is not a program. It requires a cultural change that moves the company away from a focus on events and toward processes and their improvement. It requires a move away from short-term profitability and toward short- and long-term customer satisfaction; away from hierarchies and individuals and toward teams and cooperation. Only top management can bring such a cultural change about and only by committing to the values behind such changes. Seeing this, employees, too, will take these values as their own. Finally, an important part of top management commitment is to make this new culture for quality not dependent on any one or few managers. The goal should be for the culture to become so strong that it is perpetuated by everyone in the company. (See also *Culture.*)

> For more information, see: James C. Collins and Jerry I. Porras, *Built to Last* (New York: HarperBusiness, 1994); Thomas H. Berry, *Managing for Total Quality Transformation* (New York: McGraw-Hill, 1991).

Total Customer Service A widely used term, it encompasses all the acts, features, and data that enhance a customer's ability to enjoy the potential capabilities and benefits of a service or product.

Implementing total customer service requires a careful and regular monitoring and evaluating of exactly what it is that customers consider to be good service. This changes as customers gain experience with any product and its accompanying services. You cannot relax in the quest for this type of service. That is an invitation for competitors to come in and take customers. Also, since companies exist to meet the needs of customers, and any wavering in that mission undermines the company's foundation.

> For more information, see: Valarie A. Zeithaml, A. Parasuraman, and Leonard L. Berry, *Delivering Quality Service* (New York: The Free Press, 1990).

Total Integrated Logistics This is the integration of all logistics concerning inputs into an enterprise, all processes within the organization, and the outputs of the enterprise. The purpose of looking at all the inflows and outflows of an organization is to ensure a comprehensive customer support strategy that is optimized for cost and quality.

> For more information, see: James H. Saylor, *TQM Field Manual* (New York: McGraw-Hill, 1992).

TPM aims at improving existing plant conditions and at increasing the knowledge and skills of frontline personnel in order to achieve Zero Accidents, Zero Defects, and Zero Breakdowns.

Masaji Tajiri and Fumio Gotoh

Total Productive Maintenance (TPM) This is a new approach to plant maintenance that combines productive maintenance procedures with total quality control and employee involvement to maximize the utility of productive resources. An overall goal is zero breakdowns. To achieve this, Masaji Tajiri and Fumio Gotoh (see reference) explain that TPM consists of six basic activities:

1. Elimination of the six big losses based on project teams organized by production, maintenance, and plant engineering departments.
2. Planned maintenance carried out by the maintenance department.
3. Autonomous maintenance carried out by the production department.
4. Preventive engineering carried out mainly by the plant engineering department.
5. Easy-to-manufacture product design carried out mainly by the product design department.
6. Education to support the previous activities.

For more information, see: Masaji Tajiri and Fumio Gotoh, *TPM Implementation* (New York: McGraw-Hill, 1992).

Total Quality Control (TQC) Originally conceived by quality expert Armand V. Feigenbaum, it is a system used to integrate quality development, maintenance, and improvement of all parts of an organization. The goal is to do this so that the company most economically manufactures goods and executes services while providing excellent customer service during and after the sale. In other words, it is the comprehensive set of actions organizations can take to guarantee that they efficiently deliver quality outputs to customers. It includes all the traditional quality control tools, such as inspection and sampling, as well as statistical process control.

For more information, see: Armand V. Feigenbaum, *Total Quality Control*, Third Edition, Revised (New York: McGraw-Hill, 1991).

Total Quality Management (TQM) There are several definitions for this term, but at its heart is a set of management practices designed to continuously improve the performance of organizational processes to profitably satisfy customers. It is founded on the understanding that *organizations are systems with processes that have the purpose of serving customers.*

TQM calls for the integration of all organizational activities to achieve the goal of serving customers. It seeks to impose standards, achieve efficiencies, to define roles of individuals within processes and the organization as a whole, to reduce errors and defects by applying statistical process control, and to employ teams to more efficiently plan and execute processes. It requires leaders willing to create a culture where people define their roles in terms of being responsible teammates and in terms of the value they add in delivering quality outputs to customers. Figure 79 summarizes the concerns and focuses of TQM.

For more information, see: James W. Cortada and John A. Woods, *The Quality Yearbook* (New York: McGraw-Hill, published annually); Brian L. Joiner, *Fourth Generation Management* (New York:

> The new management model is nothing more than a discipline for seeing your entire organization, the interrelationships among people and processes that determine success and the patterns of change that demand vigilance.
>
> *Stephen George and Arnold Weimerskirch*

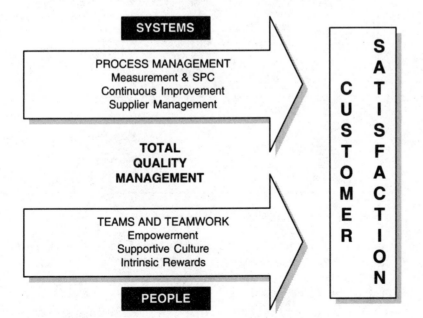

FIGURE 79

TQM involves the management of processes and the empowerment of people to create satisfied customers and a growing, profitable company.

McGraw-Hill, 1994); David A. Garvin, *Managing Quality* (New York: The Free Press, 1988); Stephen George and Arnold Weimerskirch, *Total Quality Management* (New York: John Wiley & Sons, 1994); Thomas H. Berry, *Managing the Total Quality Transformation* (New York: McGraw-Hill, 1991).

Trend This is simply a pattern evident on a control or run chart. You typically see this as points on a line chart moving generally up or generally down through time.

While the average variation as shown on a range chart may not change much, the trend line on the X-bar chart shows something is wrong. Often this is due to tool wear. For example, a grinder wheel may begin to wear out resulting in a trend of measurements up or down on a chart, but with a consistent amount of variation from the beginning, when the grinder wheel was new, throughout the process, as it begins to wear. Figure 80 shows a control chart indicating a trend at work. (See also *Drift* and *Process Drift.*)

FIGURE 80

A control chart showing a trend, indicating a specific problem somewhere in the system, such as tool wear.

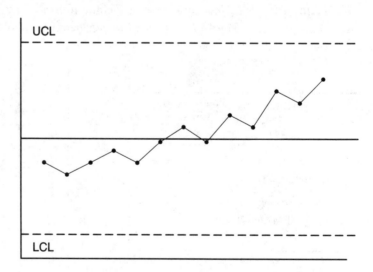

For more information, see: H.G. Menon, *TQM in New Product Manufacturing* (New York: McGraw-Hill, 1992).

Type I Error After viewing the statistical results of a sample of goods, this represents a decision that a process average and control limits have shifted in some way, based on a change from previous samples from previous lots, when they really have not. This is an important concept in manufacturing. Such errors can bring about process tampering, which can introduce additional sources of variation into a stable process. Statisticians and engineers have developed various methods for avoiding type I errors.

For more information, see: Eugene L. Grant and Richard S. Leavenworth, *Statistical Quality Control*, Sixth Edition (New York: McGraw-Hill, 1988).

Type II Error This represents a decision that a process average and control limits have not shifted from readings in samples from previous lots, when in fact they have shifted. This is also important in manufacturing. Not detecting a shift in the stability of a process will result in outputs that may not meet specifications. Statisticians and engineers have developed a variety of control charts to help avoid type II errors.

For more information, see: Eugene L. Grant and Richard S. Leavenworth, *Statistical Quality Control*, Sixth Edition (New York: McGraw-Hill, 1988).

U Chart This is a *count-per-unit chart* (see this entry).

Upper Control Limit (UCL) This is the control limit for points above the center line, the process mean, on a control chart. It is usually set at three standard deviations above the mean.

Value This is a judgment about quality that relates conformance, utility, customer satisfaction, and price. It suggests that a product with features that add more costs than utility in the view of the customer does not represent value as far as that customer is concerned. Those organizations best able to package tangible features with additional customer services, including time and place utility (see *customer delight*), at an attractive price are delivering value. Other terms that capture value are "excellence and worth" and "affordable excellence."

> For more information, see: David A. Garvin, *Managing Quality* (New York: The Free Press, 1988); H. James Harrington, *Business Process Improvement* (New York: McGraw-Hill, 1991).

For most business processes, less than 30 percent of the *cost* is in value-added activities. Even more alarming is the mismatch of cycle of value-added activities compared to total cycle time. For most business processes, less than 5 percent of time is spent in value-added activities.

H. James Harrington

Value-Added Activities These are organizational activities directly involved in converting inputs into outputs that meet customer needs and expectations and will generate customer satisfaction. We can contrast value-added activities with activities that do not add value but do add costs. Many organizations have, over time, added many different activities to their processes to cover contingencies and check the checkers. There was always a reason for adding such activities, and over time they become institutionalized, with no one really questioning their utility or validity. However, when companies begin to institute quality management and take a close look at their processes, they always find activities that are not necessary or directly involved in delivering outputs to customers. These are targets for elimination. Figure 81 shows a process flowchart for analyzing activities to determine whether they add costs or value to final outputs.

FIGURE 81

A process for identifying activities that add value and those that just add costs. Adapted from H. James Harrington, *Business Process Improvement* (New York: McGraw-Hill, 1991) p. 141.

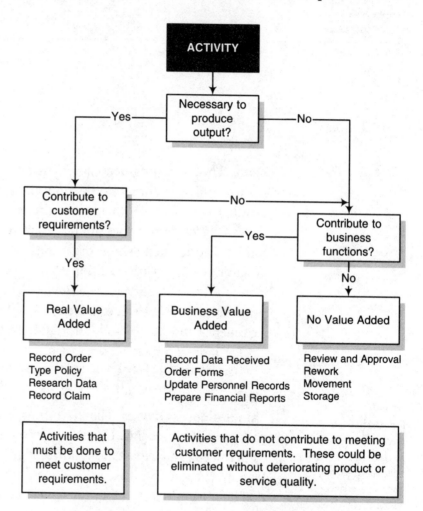

For more information, see: H. James Harrington, *Business Process Improvement* (New York: McGraw-Hill, 1991); Brian L. Joiner, *Fourth Generation Management* (New York: McGraw-Hill, 1994); Thomas H. Berry, *Managing the Total Quality Transformation* (New York: McGraw-Hill, 1991).

Value Chain This represents the chain of activities and processes by which value is added to inputs resulting in the delivery of products and services to final customers. By reviewing the value chain, you can identify which activities add value and which add costs. Value chains can be helpful in improving processes because they show steps that are critical and those that a company might be able to eliminate. Figure 82 illustrates the value chain for delivering this book to final customers.

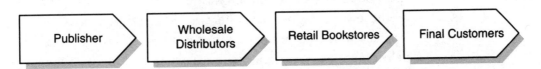

For more information, see: Brian L. Joiner, *Fourth Generation Management* (New York: McGraw-Hill, 1994); Michael E. Porter, *Competitive Advantage* (New York: The Free Press, 1985).

FIGURE 82

A simple value chain for the delivery of books to final customers. Each step in this chain adds some value to getting a product to customers.

Value Engineering This is a U.S. Department of Defense concept. DoD defines it as follows: It is an organized effort directed at analyzing the function of systems, equipment, facilities, service, and supplies for the purpose of achieving essential functions at the lowest life cycle cost consistent with performance, reliability, maintainability, interchangeability, product quality, and safety. In other words, it is an approach that focuses engineers on improving productivity and quality while reducing costs. It is another way of talking about Total Quality Management and all the activities that covers.

For more information, see: James H. Saylor, *TQM Field Manual* (New York: McGraw-Hill, 1992).

Variable A variable is one piece of measurable data from a larger set that falls within some range with a certain frequency or pattern. A variable is always measurable, such as height, weight, temperature, time, and so on. (See also *Attribute.*)

For more information, see: George L. Miller and LaRue L. Krumm, *The Whats, Whys & Hows of Quality Improvement* (Milwaukee, WI: ASQC Quality Press, 1992); Hy Pitt, *SPC for the Rest of Us* (Reading, MA: Addison-Wesley, 1994).

Variance This is nonconformance to expectations or specifications. In statistics, variance is the square of the standard deviation. In quality, it is simply how far off the mark something is from meeting customer expectations.

In traditional management practice, dealing with variance can get a company in trouble. Frequently reports are issued that show budget targets and then the actual numbers with variance as the difference between the two. Sometimes managers are tempted to act when one number, such as sales for a month, is below target or at variance with expectations. Unless a manager knows whether this result is within the range of control limits or due to a special cause, it can lead to tampering, which just adds more uncertainty to the process.

For more information, see: George L. Miller and LaRue L. Krumm, *The Whats, Whys & Hows of Quality Improvement* (Milwaukee, WI: ASQC Quality Press, 1992); Hy Pitt, *SPC for the Rest of Us* (Reading, MA: Addison-Wesley, 1994); Brian L. Joiner, *Fourth Generation Management* (New York: McGraw-Hill, 1994).

Variation This is an idea that suggests there will always be some difference between any two or more actions within a system and in the outputs of that system. In systems, there is always continuous interaction going on among components that affects the output of that system. That output variation can be great or small depending on how well the system is operating.

> The central problem in management and in leadership is failure to understand the information in variation.
>
> *W. Edwards Deming*

With an in-control system, you can determine the amount of expected variation from measurement to measurement. It will fall within the upper and lower control limits of the process. The only way to decrease variation with an in-control system is to address the system as a whole, looking at interactions that can be refined. Conversely, when a variation falls outside control limits, you know it is not inherent in the system, and you can immediately look for and do something about its cause. This understanding of systems and variation is an important part of TQM.

For more information, see: W. Edwards Deming, *The New Economics* (Cambridge, MA: MIT Center for Advanced Engineering Study, 1993); Brian L. Joiner, *Fourth Generation Management* (New York: McGraw-Hill, 1994); George L. Miller and LaRue L. Krumm, *The Whats, Whys & Hows of Quality Improvement* (Milwaukee, WI: ASQC Quality Press, 1992).

Vision This is, according to Burt Nanus, "a realistic, credible, attractive future for your organization. It is your articulation of a destination toward which your organization should aim, a future that in important ways is better, more successful, or more desirable." We can contrast vision with mission, which articulates the purpose of a company. Vision-based strategic planning is then focused less on today's problems and more on what strategies are required to make the vision come true.

Vision is important because it provides direction for everyone in the company. It tells them what they are working toward. In a quality culture, one goal is to have everyone contribute to and share the vision of top management. In this way, they are more likely to be committed to this vision and to take actions to help make it a reality.

For more information, see: Burt Nanus, *Visionary Leadership* (San Francisco, Jossey-Bass, 1992); Peter M. Senge, *The Fifth Discipline* (New York: Doubleday Currency, 1990); Ernst & Young, *Total Quality: An Executive's Guide for the 1990s* (Burr Ridge, IL: Irwin Professional Publishing, 1990).

Vital Few, Trivial Many Joseph M. Juran coined this phrase to describe use of the Pareto concept in which most effects come from a few causes, that is, 80 percent of effects from 20 percent of possible causes. The 20 percent of possible causes are the "vital few," the other causes are the "trivial many."

For more information, see: Joseph M. Juran, *Managerial Breakthrough*, Revised Edition (New York: McGraw-Hill, 1994).

Voice of the Customer This is a basic concept in quality management and signifies efforts to develop processes to capture what customers are saying by their words and behaviors about an organization's products and services.

Listening to the voice of the customer is absolutely necessary (but not sufficient) to remain competitive. There are a variety of techniques for capturing what customers have to tell you. Some of these methods include traditional market research, including surveys, focus groups, feedback forms, and data from scanners and other sources. Another important method involves being able to capture customer complaints, either to customer service representatives or to any employees who deal with customers directly. Many companies look on complaints as individual problems to solve and never look at this as information that can guide improvements and improve customer satisfaction.

While listening to the voice of the customer can help a company improve its products and services and can also provide ideas for new products, companies cannot depend on customers to tell researchers what they want in the future. Deming liked to say that no customer ever asked for a lightbulb. What they do request is better and more reliable light sources. Figuring out how to address that is still the province of creative employees seeking to deliver benefits packaged as products and services that customers will find valuable. Finally, one important tool for incorporating the voice of the customer into products is the *house of quality* and *quality function deployment* in new product planning (see these terms).

For more information, see: Richard C. Whiteley, *The Customer-Driven Company* (Reading, MA: Addison-Wesley, 1992); Brian L. Joiner, *Fourth Generation Management* (New York: McGraw-Hill, 1994); Sarv Singh Soin, *Total Quality Control Essentials* (New York: McGraw-Hill, 1992).

Voice of the Process This is the actual nature of the output of a process in terms of variation, costs, and quality. In other words, it is the result that the process gives you.

A process may or may not operate within specifications and may or may not deliver outputs that meet a company's expectations or those of its customers. Nevertheless, only by listening to the voice of the process through the use of control charts and other SPC tools and techniques can you begin to understand its capabilities. The goal for any organization is to match the voice of its processes with the voice of its customers.

For more information, see: William Scherkenbach, *Deming's Road to Continual Improvement* (Knoxville, TN: SPC Press, 1991).

World-Class This descriptive term is often used with such expressions as world-class quality, world-class company, world-class process. It means the best in the world. It is often used in benchmarking activities. In terms of the Malcolm Baldrige Award Criteria, world-class has come to mean a firm that has achieved at least 876 points out of 1,000 in its quality appraisal.

In benchmarking, companies may look for best-of-class operations, which they often locate within their own industry, or they may look for world-class, which takes in any company in any industry. (See also *Benchmarking*.)

For more information, see: H. James Harrington, *Business Process Improvement* (New York: McGraw-Hill, 1991); Michael J. Spendolini, *The Benchmarking Book* (New York: Amacom, 1992).

X-Bar Chart This is the formal term for *average chart* (see this term for an example).

Zero Defects This is an aspiration but also a way of explaining to employees the notion that everything should be done right the first time, that there should be no failures or defects in work outputs. The idea was popularized by Philip B. Crosby, and it forms the basis of his idea of quality. It is somewhat controversial because some quality experts feel it mainly focuses on internally meeting design specifications at every step in a process and delivering defect-free outputs. It does not place a heavy focus on customers nor on continuous improvement.

For more information, see: Philip B. Crosby, *Quality Is Free* (New York: McGraw-Hill, 1979).

APPENDIX 1

QUALITY REFERENCES

Throughout this book we have made references to a wide variety of titles where you can find additional information on the many terms and concepts we've selected for inclusion. In this appendix, we want to summarize that list of books by broad topic area. The categories we have chosen to do this include the following:

- Sampling, Quality Control, and SPC
- Quality in Manufacturing
- Cycle Time Reduction
- Total Quality Management
- Leadership, People, and Culture
- Teams and Teamwork
- Benchmarking
- Process Improvement and Reengineering
- Baldrige Assessments and ISO 9000
- Customer Focus

This should be a convenient way for you to consider books that may be appropriate for your library. Some of the books in this listing are technical in nature and mainly appropriate for engineers and technicians. This listing is not meant to be exhaustive, but it includes both enough variety and enough different titles to get you going on any subject in which you are interested. We include a few remarks about each book to help you choose among them.

SAMPLING, QUALITY CONTROL, AND SPC

This section lists titles that explain the use of statistics in quality control and management activities. Some are technical, but most are accessible to anyone interested in the subject.

Michael Brassard and Diane Ritter, *The Memory Jogger II* (Methuen, MA: GOAL/APC, 1994).

This is a small book for the shirt pocket or purse that reviews the basic tools of quality management. Widely used.

Armand V. Feigenbaum, *Total Quality Control,* Third Edition, Revised (New York: McGraw-Hill, 1991).

This book is somewhat technical but it is highly recommended as a reference on nearly any topic having to do with sampling plans and statistical process control in general. It is understandable by most nontechnicians who have an interest in a particular issue.

Howard Gitlow, Shelly Gitlow, Alan Oppenheim, and Russ Oppenheim, *Tools and Methods for the Improvement of Quality* (Burr Ridge, IL: Irwin, 1989).

A good basic textbook on these issues. Not too technical for those willing to expend a little effort.

Eugene L. Grant and Richard S. Leavenworth, *Statistical Quality Control,* Sixth Edition (New York: McGraw-Hill, 1988).

This is a standard textbook on these issues and can serve as a reference as well. Fairly technical.

John L. Hradesky, *Productivity & Quality Improvement: A Practical Guide to Implementing Statistical Process Control* (New York: McGraw-Hill, 1988).

A clear straightforward book for those directly involved in SPC on a daily basis. Lots of good how-to material included. Recommended.

Kaoru Ishikawa, *Introduction to Quality Control* (Tokyo, 3A Corporation, 1990).

Ishikawa is a pioneer in quality control techniques, and this book presents his ideas on this subject. Somewhat technical.

Joseph M. Juran and Frank M. Gryna, *Quality Planning and Analysis*, Third Edition (New York: McGraw-Hill, 1993).

A comprehensive title covering many of the subjects found throughout the encyclopedia. If you were only going to purchase three references from our entire listing here, this should be one of them. Highly recommended.

Henry Lefevre, *Quality Service Pays: Six Keys to Success* (Milwaukee, WI: ASQC Quality Press, 1989).

This title emphasizes the use of SPC techniques in managing service activities. One of the few titles oriented toward services rather than manufacturing. Somewhat technical, but for the most part understandable. Good coverage of quality tools.

N. Logothetis, *Managing For Total Quality: From Deming to Taguchi and SPC* (Englewood Cliffs, NJ: Prentice Hall, 1992).

Published originally in the U.K., this is a comprehensive review of SPC tools and techniques. Somewhat technical, but recommended because of many topics.

Eugene H. Melan, *Process Management: Methods for Improving Products and Services* (New York: McGraw-Hill, 1993).

A good, sound book for understanding basic process management techniques.

H.G. Menon, *TQM in New Product Manufacturing* (New York: McGraw-Hill, 1992).

Short but thorough introduction to the use of quality management and control techniques in manufacturing.

George L. Miller and LaRue L. Krumm, *The Whats, Whys & Hows of Quality Improvement* (Milwaukee: WI: ASQC Quality Press, 1992).

This covers a variety of topics on sampling, quality control, and SPC. It is written for the uninitiated and is quite understandable. Good coverage of quality tools and techniques. Recommended for nontechnicians.

Hy Pitt, *SPC for the Rest of Us* (Reading, MA: Addison-Wesley, 1994).

Like its title says, this is a down-to-earth book on quality control issues that is quite understandable for the beginner. Highly recommended.

Ellis R. Ott and Edward G. Schilling, *Process Quality Control*, Second Edition (New York: McGraw-Hill, 1990).

This is a classic on sampling theory and quality control. Very technical.

Phillip J. Ross, *Taguchi Techniques for Quality Engineering: Loss Function, Orthogonal Experiments, Parameter Design, and Tolerance Design* (New York: McGraw-Hill, 1988).

The subtitle says what it covers. The Taguchi approach is based on minimizing variance from targets and provides many techniques for doing this. Somewhat technical.

J.R. Taylor, *Quality Control Systems* (New York: McGraw-Hill, 1989).

A book with a lot of lists for setting up quality control systems in a variety of manufacturing environments.

Donald J. Wheeler and David S. Chambers, *Understanding Statistical Process Control*, Second Edition (Knoxville, TN: SPC Press, 1992).

This is a textbook, but less technical than some others. It can serve as a useful reference for the novice.

Quality in Manufacturing

B. Joseph Pine II, *Mass Customization: The New Frontier in Business Competition* (Cambridge, MA: Harvard Business School Press, 1993).

This is the pioneering book on the subject of flexibility in manufacturing to deliver maximum customer satisfaction. Recommended.

Dev G. Raheja, *Assurance Technologies: Principles and Practices* (New York: McGraw-Hill, 1991).

This book is aimed at engineers and includes material on reliability, maintainability, safety, and quality assurance. Somewhat technical.

Masaji Tajiri and Fumio Gotoh, *TPM Implementation: A Japanese Approach* (New York: McGraw-Hill, 1992).

TPM means Total Productive Maintenance. The Japanese approach combines productive maintenance with total quality control techniques and employee involvement. It's all about keeping the plant running with minimum breakdowns.

See also the titles under *Sampling, Quality Control, and SPC* by Feigenbaum, Hradesky, Juran & Gryna, Melan, Menon, Ross, and Taylor, which deal with quality manufacturing issues.

Cycle Time Reduction

Gerard H. Gaynor, *Exploiting Cycle Time in Technology* (New York: McGraw-Hill, 1993).

A complete review of cycle time reduction issues aimed mainly at technicians and engineers responsible for implementing it. Understandable by all—no numbers.

George Stalk, Jr. and Thomas M. Hout, *Competing Against Time: How Time-Based Competition is Reshaping Global Markets* (New York: The Free Press, 1990).

The authors argue that doing things faster is the best competitive strategy today. Their case is powerful and their prescriptions practical.

Total Quality Management

The titles listed here are all about implementing what has come to be known as TQM in any organization—profit, nonprofit, manufacturing, services, retailing, wholesaling, government, education, or health care. The principles of TQM apply equally to all. This list includes classics by Deming and Juran as well as a variety of contemporary titles we think are worth your consideration.

Thomas H. Berry, *Managing the Total Quality Transformation* (New York: McGraw-Hill, 1991).

An excellent book for beginners. It provides a good review of the basics in an accessible style. Recommended.

Greg Bounds, Lyle Yorks, Mel Adams, Gipsie Ranney, *Beyond Total Quality Management: Toward the Emerging Paradigm* (New York: McGraw-Hill, 1994).

This is a college textbook, one of the first, on the principles of TQM and their implementation in organizations. It is comprehensive and can be a useful reference.

Bruce Brocka and M. Suzanne Brocka, *Quality Management: Implementing the Best Ideas of the Masters* (Burr Ridge, IL: Irwin Professional Publishing, 1992).

A comprehensive review of quality tools, techniques, and principles. Appropriate for anyone. A good reference. Recommended.

James W. Cortada and John A. Woods, *The Quality Yearbook* (New York: McGraw-Hill, published annually). Call 1-800-262-4729 for information.

This is the only annual anthology and reference to TQM available. It includes current articles from many publications on every aspect of quality management. The reference section includes a bibliography of books and articles published in the current year, a review of quality awards, journals and magazines, current Baldrige Award winners, and much other useful information.

Philip B. Crosby, *Quality Is Free: The Art of Making Quality Certain* (New York: McGraw-Hill, 1979).

This is the book that started people thinking about zero defects and that focusing on quality lowers costs rather than raises them.

W. Edwards Deming, *Out of the Crisis* (Cambridge, MA: MIT Center for Advanced Engineering Study, 1986).

Deming's summary of his approach to management. It presents a comprehensive review of his 14 Points for Management and the 7 Deadly Diseases. It also includes his ideas on understanding process control and the implications of that for managing successfully. Recommended for anyone interested in quality management, though it is written in an idiosyncratic style.

W. Edwards Deming, *The New Economics: For Industry, Government, Education* (Cambridge, MA: MIT Center for Advanced Engineering Study, 1993).

This was the last book Dr. Deming published before his death in December 1993. It gives you a sample of his thinking on all his key ideas. It includes his ideas on profound knowledge and explains the Red Beads Experiment.

David A. Garvin, *Managing Quality: The Strategic and Competitive Edge* (New York: The Free Press, 1988).

Garvin, a Harvard Business School professor, provides an excellent overview of quality management practices and their strategic importance in establishing and maintaining a competitive advantage. Recommended.

Stephen George and Arnold Weimerskirch, *Total Quality Management: Strategies and Techniques Proven at Today's Most Successful Companies* (New York: John Wiley & Sons, 1994).

Part of the Portable MBA Series from Wiley, this book does a good job of covering the basics with examples of how prominent companies do it.

Masaaki Imai, *KAIZEN: The Key to Japan's Competitive Success* (New York: McGraw-Hill, 1986).

Well-written and insightful. The principles of Kaizen (continuous improvement) are very applicable to American business. Recommended.

Brian L. Joiner, *Fourth Generation Management: The New Business Consciousness* (New York: McGraw-Hill, 1994).

A complete review of the fundamental ideas of quality management—customer focus, process management, and teamwork. It includes a very understandable discussion of the use of control charts in all organizational activities, including why they are necessary and why they work. Highly recommended.

Joseph M. Juran, *Juran on Leadership for Quality: An Executive Handbook* (New York: The Free Press, 1989).

This quality guru has published many important books on the subject. This one is an excellent introduction to his ideas. If you could read only one of his books, this one would be our recommendation.

Joseph M. Juran, *Managerial Breakthrough*, Revised Edition (New York: McGraw-Hill, 1994).

When originally published in 1964, this book was way ahead of its time. The revised 1994 edition includes sound advice on the use of quality management techniques in any organization.

William Lareau, *American Samurai* (New York: Warner Books, 1991).

This is an excellent and comprehensive introduction to quality management ideas. It is frequently referenced in the encyclopedia. One of our favorite books, it is written in an informal style and provides any manager with a thorough grounding in the why and how of implementing TQM. Highly recommended.

Marshall Sashkin and Kenneth J. Kiser, *Putting Total Quality Management to Work* (San Francisco: Barrett-Koehler Publishers, 1993).

A brief but sound introduction to quality management. Especially valuable for its coverage of the culture of quality.

Saylor, James H. *TQM Field Manual*, (New York: McGraw-Hill, 1992).

Not profound, but lots of practical advice about tools and techniques written in a straightforward and easily accessible manner. Recommended.

William W. Scherkenbach, *Deming's Road to Continual Improvement* (Knoxville, TN: SPC Press, 1991).

This book is aimed at explaining Deming's approach to managing for quality and does a good job explaining process management and making the change to a process orientation. Scherkenbach was a teacher at Deming's four-day seminar.

Sarv Singh Soin, *Total Quality Control Essentials: Key Elements, Methodologies, and Managing for Success* (New York: McGraw-Hill, 1992).

This is a nice blend of basic quality management principles with coverage of some important tools, including good coverage of quality function deployment and overall process management. Recommended.

Roger Tunks, *Fast Track to Quality: A 12-Month Program for Small to Mid-Sized Businesses* (New York: McGraw-Hill, 1992).

Tunks does a good job of introducing readers to all the ideas involved in quality management with specific advice on implenting TQM in terms of culture and actions.

Mary Walton, *The Deming Management Method* (New York: Perigee Books, 1986).

This is the standard work for any beginner to Deming's methods. It carefully explains the 14 Points and 7 Deadly Diseases and provides several examples of these principles at work. Recommended.

Leadership, People and Culture

The books here deal with leadership and people issues involved in implementing quality management in any organization—soft-side stuff, but very important.

James C. Collins and Jerry I. Porras, *Built to Last: Successful Habits of Visionary Companies* (New York: HarperBusiness, 1994).

This book reveals the results of studies of great companies, showing that rather than depending on charismatic leaders, they have great cultures that bring out the best in everyone. Highly recommended.

Max DePree, *Leadership Is an Art* and *Leadership Jazz* (New York: Doubleday, 1989 and 1992).

Both of these books include a series of short essays on leadership designed to provoke thinking and inspire readers. At the heart of these books is the idea that leadership is caring about your people and getting them to care about themselves and their colleagues.

Craig R. Hickman, *Mind of a Manager, Soul of a Leader* (New York: John Wiley & Sons, 1990).

In a series of 49 short chapters, the author contrasts managerial action with leadership action, to bring out the best in people and organizations. Recommended.

Alfie Kohn, *No Contest: The Case Against Competition*, Revised Edition (Boston: Houghton Mifflin, 1992).

This provocative book discusses the destructive nature of competition among individuals and makes a strong case for attitudes and cultures that promote cooperation, which is necessary for successfully implementing TQM. Recommended.

Alfie Kohn, *Punished by Rewards: The Trouble with Gold Stars, Incentive Plans, A's Praise, and Other Bribes* (Boston, Houghton Mifflin, 1993).

This is Kohn's other provocative and controversial book on rewards. His important point is that external rewards undermine intrinsic motivation and help bring about mediocrity. This idea has value in implementing TQM, which works best in cultures that allow intrinsic motivation to blossom. Recommended.

Burt Nanus, *Visionary Leadership: Creating a Compelling Sense of Direction for Your Organization* (San Francisco: Jossey-Bass, 1992).

This book is all about how leaders create a vision for their organizations and then implement it.

Harry V. Roberts and Bernard E. Sergesketter, *Quality Is Personal* (New York: The Free Press, 1993).

The authors explain how to use the principles of quality for personal improvement. Practical and recommended.

Robert H. Rosen, *The Healthy Company: Eight Strategies to Develop People, Productivity, and Profits* (Los Angeles: Jeremy P. Tarcher/Perigee, 1991).

While not dealing with quality management per se, it provides many ideas for developing an environment in which employees can maximize their contributions to the organization.

Peter M. Senge, *The Fifth Discipline: The Art and Practice of the Learning Organization* (New York: Doubleday Currency, 1990).

This is a great book on understanding organizations as systems and how to act on this understanding to create a great company. Because

TQM requires the systems view to truly work, this book provides an excellent sense of what that means. If you only have time to read two or three books, this should be one of them. Highly recommended.

Frank K. Sonnenberg, *Managing With a Conscience: How to Improve Performance Through Integrity, Trust, and Commitment* **(New York: McGraw-Hill, 1994).**

This book is about developing an organization in which integrity and respect for individuals prevails, not just because this seems like a good idea, but because it is the best way to get results.

John O. Whitney, *The Trust Factor: Liberating Profits and Restoring Corporate Vitality* **(New York: McGraw-Hill, 1994).**

Like Sonnenberg's, this book looks at how to develop an environment of trust and what this means for competing successfully.

Teams and Teamwork

Kimball Fisher, Steven Rayner, and William Belgard, *Tips for Teams: A Ready Reference for Solving Common Team Problems* **(New York: McGraw-Hill, 1994).**

Just what its title says, this is a paperback reference for facilitating teamwork.

Charles C. Manz and Henry P. Sims, Jr., *Business Without Bosses: How Self-Managing Teams are Building High-Performing Companies* **(New York: John Wiley & Sons, 1993).**

This is a practical guide to implementing self-managed teams. It includes many interesting examples from real companies. Recommended.

Jack D. Orsburn, Linda Moran, Ed Musselwhite, and John H. Zenger, *Self-Directed Work Teams: The New American Challenge* **(Burr Ridge, IL: Irwin Professional Publishing, 1990).**

This is a lucid account of how to create, deploy, and manage teams in a business environment. While there are many such books available, this one has stood the test of time as a practical guide to tactical implementation.

Peter R. Scholtes, *The Team Handbook* **(Madison, WI: Joiner Associates, 1988).**

A perennial bestseller, this is the leading title for training people to work as teams. It includes lots of background on quality practices and tools, lots on teamwork, and a wealth of how-to exercises. It is frequently referenced in this encyclopedia. Highly recommended.

Benchmarking

Benchmarking, comparing your practices with those of the best companies, requires background and preparation to figure how to do it and to use what you learn. These are the books we referred to in preparing this encyclopedia.

Christopher E. Bogan and Michael J. English, *Benchmarking for Best Practices: Winning Through Innovative Adaptation* **(New York: McGraw-Hill, 1994).**

This is a comprehensive guide to benchmarking that tells you everything you want to know about this subject. Recommended.

Robert J. Boxwell, Jr., *Benchmarking for Competitive Advantage* **(New York: McGraw-Hill, 1994).**

This book is a practical guide that shows how to plan and execute a benchmarking study and develop an action plan based on results. Its appendices of organizations and companies is especially valuable.

Robert C. Camp, *Benchmarking: The Search for Industry Best Practices That Lead to Superior Performance* **(Milwaukee, WI: ASQC Quality Press, 1989).**

This is the first and most often cited text on benchmarking. Camp developed the benchmarking program at Xerox. Excellent on tactical implementation issues.

Michael J. Spendolini, *The Benchmarking Book* **(New York: Amacom, 1992).**

Spendolini is a consultant whose practice focuses on benchmarking. A good basic review of this subject.

Gregory H. Watson, *Strategic Benchmarking: How to Rate Your Company's Performance Against the World's Best* (New York: John Wiley & Sons, 1993).

This book focuses on the strategic issues surrounding benchmarking and how to benchmark one company's strategy with another. Recommended.

Process Improvement and Reengineering

This section looks at some basic how-to books for continuous improvement and reengineering of processes.

Thomas H. Davenport, *Process Innovation: Reengineering Work Through Information Technology* (Cambridge, MA: Harvard Business School Press, 1993).

Davenport takes you through a practical step-by-step approach on applying computer technology to process reengineering. It is the best book on computers and quality available today.

Michael Hammer and James Champy, *Reengineering the Corporation: A Manifesto for Business Revolution* (New York: HarperBusiness, 1993).

The book that popularized reengineering. The authors are the most widely recognized proponents of radically redesigning major corporate processes, rather than simply improving them continuously. They state their case, show how it is done, and conclude with examples.

H. James Harrington, *Business Process Improvement* (New York: McGraw-Hill, 1991).

Harrington's book became the overnight standard work on how to improve a process. It is a step-by-step guide. If you could only read one how-to book on process improvement, this would be it. Frequently referenced in the encyclopedia. Highly recommended.

H. James Harrington, *Total Improvement Management: Creating the Custom-Tailored Turnaround Strategy* (New York: McGraw-Hill, 1994).

This is Harrington's comprehensive guide for managing with continuous improvement in mind. Includes a 15-step program for implementing this approach to management.

Raymond L. Manganelli and Mark M. Klein, *The Reengineering Handbook: A Step-By-Step Guide to Business Transformation* (New York: Amacom, 1994).

Just what its title implies, this book provides step-by-step directions for implementing reengineering projects.

Daniel Morris and Joel Brandon, *Re-engineering Your Business* (New York: McGraw-Hill, 1993).

This explains what it is and how to do it in areas like business processes, information technology, and human resources.

Baldrige Assessments and ISO 9000

These books can be valuable in using the Baldrige Criteria to assess current implementation of quality management or in preparing for ISO 9000 certification.

Donald C. Fisher, *Measuring Up to the Baldrige: A Quick and Easy Self-Assessment Guide for Organizations of All Sizes* (New York: Amacom: 1994).

The subtitle says it all. A good how-to book for using the Baldrige Criteria to figure out where you stand.

Christopher W.L. Hart and Christopher E. Bogan, *The Baldrige: What It Is, How It's Won, How To Use It To Improve Quality in Your Company* (New York: McGraw-Hill, 1992).

Again, the subtitle says it all. It systematically goes through the criteria and how to use them to make self-assessments.

Charles A. Mills, *The Quality Audit: A Management Education Tool* (New York: McGraw-Hill, 1989).

This is an excellent reference on planning and executing quality audits. It includes a comprehensive bibliography of books and articles. Useful for understanding ISO 9000 certification.

John T. Rabbit and Peter A. Bergh, *The ISO 9000 Book,* Second Edition (New York: Amacom, 1994).

An easy-to-understand guide to ISO 9000 certification. Takes the mystery out of this process. Recommended.

Customer Focus

Here we review some titles that provide insight into understanding customer requirements and managing the organization to deliver the goods.

James W. Cortada, *TQM For Sales and Marketing Management* (New York: McGraw-Hill, 1993).

This is one of the few books available that systematically applies quality management practices to the sales and marketing process.

Theodore Levitt, *Marketing For Business Growth* (New York: McGraw-Hill, 1974).

This book is probably out of print, but if you can find it, it is worth the effort. Levitt provides an insightful understanding of what it is that customers value (benefits) and how to make that idea the driver of management decisions. Highly recommended.

Hal F. Rosenbluth, *The Customer Comes Second: And Other Secrets of Exceptional Service* (New York: Quill/William Morrow, 1992).

This is an excellent first person account of how one travel agency created a culture where quality management practices thrive and, by its way of operating, dramatically improved its sales to over $1.5 billion dollars annually. Recommended.

Eberhard E. Scheuing and William F. Christopher, *The Service Quality Handbook* (New York: Amacom, 1993).

This includes 57 original contributions from experts on everything you ever wanted to know about service and imbuing it with quality. Recommended.

Richard J. Schonberger, *Building a Chain of Customers: Linking Business Functions to Create the World Class Company* (New York: The Free Press, 1990).

This book examines how companies can look at processes as satisfying the needs and wants of a chain of customers, each adding value to what the company delivers to final customers. An influential book.

Richard C. Whiteley, *The Customer-Driven Company: Moving From Talk to Action* (Reading, MA: Addison-Wesley, 1991).

This is a how-to guide for using various tools and techniques to bring the customer into the company and efficiently deliver products and services that meet and exceed expectations. Recommended.

APPENDIX 2

MAGAZINES AND JOURNALS ON QUALITY MANAGEMENT

This appendix gives you several magazines that deal directly with issues surrounding quality management. We have organized this list by categories to make it easier to use. Each magazine has a rating of either recommended or highly recommended to help you judge among them. All are of interest and can serve as valuable resources.

Magazines and Journals on General Management Issues

Industry Week
Penton Publishing Company
1100 Superior Avenue
Cleveland, OH 44114-2543
(216) 696-7000
Published biweekly.
Subscriptions: Distributed as a closed circulation magazine to qualified executives in administration, finance, production, engineering, purchasing, marketing, and sales. To those who do not qualify, it is available by subscription at $60 per year or $100 for two years; Canada $90 per year or $150 for two years; international $110 per year or $195 for two years.

Includes articles, columns, and reviews of timely interest to managers on a wide range of business and management topics, with frequent articles on quality management. Highly recommended.

Journal for Quality and Participation
Association for Quality and Participation
801-B West 8th Street, Suite 501
Cincinnati, OH 45203
(513) 381-1959
Published seven times a year (Jan/Feb, March, June, July/Aug. Sept., Oct./Nov., and Dec.).

Subscriptions: Available as part of membership in AQP. Non-members: $52 per year in the United States and $75 for international orders.

Includes a variety of practical, detailed articles on implementing quality in different industries and organizations. Articles are nearly always intriguing, useful, and well written. One of the best magazines available on quality for managers. Highly recommended.

Management Review
American Management Association
Publications Division
Box 408
Saranac, NY 12983-0408
Published monthly.
Subscriptions: $45 per year in the United States; non-United States subscriptions $60 per year.

The official magazine of the American Management Association, it includes a variety of articles of interest to managers and often includes items on quality issues. Recommended.

National Productivity Review
John Wiley & Sons, Inc.
605 Third Avenue
New York, NY 10158
(212) 592-6479
Published quarterly.
Subscriptions: United States and Canada, one year $168, two years $309. International, one year $218. Discounts available on multiple copy subscriptions.

A journal that includes practical articles focusing on the implementation of quality in all types of organizations. Divided into three sections: Ideas and Opinions, Features, and Reviews. Highly recommended.

Personnel Journal
ACC Communications
245 Fischer Avenue, B-2
Costa Mesa, CA 92626
Published monthly.
Subscriptions: $55 per year. For multiyear and international subscriptions, write to the publishers.

An exceptionally well done journal that often includes articles related to quality issues in the area of human resources, including teams and compensation. Regularly includes articles profiling practices at various companies. Highly recommended.

Quality Digest
QCI International
1350 Vista Way, P.O. Box 882
Red Bluff, CA 96080
(916) 527-6070, (800) 527-8875
Published monthly.
Subscriptions: Official rate $75 per year but discounted to $45 for United States and $69 for international orders.

Includes a variety of how-to articles and pieces on how various organizations implement quality, especially in people management. Also includes monthly columnists such as Tom Peters, Karl Albrecht, and Ken Blanchard, plus book reviews and other information. Highly recommended.

Quality Management Journal
American Society for Quality Control
611 East Wisconsin Avenue, P.O. Box 3005
Milwaukee, WI 53201-3005
(414) 272-8575, (800) 248-1946
Published quarterly.
Subscriptions: Available to members of ASQC in the United States at $50 annually, to nonmembers at $60; Canada $74 to members and $84 to nonmembers; international $74 to members and $84 to nonmembers.

A peer-reviewed journal designed to present academic research on quality management in a style that makes it accessible to managers in all fields. Recommended.

Quality Progress
American Society for Quality Control, Inc.
611 East Wisconsin Avenue, P.O. Box 3005
Milwaukee, WI 53201-3005
(414) 272-8575, (800) 248-1946
Published monthly.
Subscriptions: Available as part of membership in the ASQC. $50 per year for nonmembers in the United States and $85 for first class to Canada and international airmail.

This is the foremost magazine on quality management available. It includes a wide variety of general and technical articles plus event calendars, reviews, and many other features. Highly recommended.

Sloan Management Review
MIT Sloan School of Management
292 Main Street, E38-120
Cambridge, MA 02139
(617) 253-7170
Published quarterly.
Subscriptions: $59 per year; $79 for Canada and Mexico, and $89 for all other international subscriptions.

Includes practical, yet thoughtful articles on a variety of issues of direct interest to managers. Articles often directly or indirectly related to quality issues. Recommended.

Target
Association for Manufacturing Excellence
380 West Palatine Road
Wheeling, IL 60090
(708) 520-3282
Published bimonthly.
Subscriptions: Available to members of AME as part of membership. Cost of annual membership: $125.

Includes many practical articles on how various industries and companies are implementing TQM in manufacturing and management. Also includes reports from regional chapters, book reviews, and event calendar. Practical, accessible, and well done. Recommended.

The Quality Observer
The Quality Observer Corporation
3505 Old Lee Highway, P.O. Box 1111
Fairfax, VA 22030
(703) 691-9295
Published monthly.
Subscriptions: one year $53, two years $90, and three years $130; overseas one year $68, two years $120, and three years $175; libraries one year $90 (two copies each issues sent to same address); corporations one year $180 (Five copies each issue sent to same address).

Billed as the "International News Magazine of Quality," this is a three-color tabloid-sized publication, with case studies, international news, interviews, and regular columns on quality topics. Recommended.

The Lakewood Report
Lakewood Publications
50 South Ninth Street
Minneapolis, MN 55402
(612) 333-0471, (800) 328-4329
Published monthly.
Subscriptions: One year United States $147; Canada add $100 plus 7 percent GST; other countries add $20.

This is a twelve-page, two-color newsletter covering specific issues in the delivery of quality to customers and creating company environments where service and quality thrive for everyone. Lots of practical ideas. Recommended.

The Total Quality Review
Cambridge Strategy Publications
P.O. Box 26007
Alexandria, VA 22313-6007.
(800) 599-9001
Published bimonthly.
Subscriptions: Introductory annual subscription in the United States available at $135 (regular rate thereafter $165); international introductory annual subscription $165 (regular rate thereafter $195).

Organized around different themes each month, it includes a variety of how-to articles, case studies, and background pieces for understanding the implementation of TQM. Also includes regular news from the Council for Continuous Improvement. Recommended.

Tom Peters On Achieving Excellence
TPG Communications
P.O. Box 2189
Berkeley, CA 94702-0189
(800) 959-1059
Published monthly.
Subscriptions: United States one year $197 and two years $244. Call for international rates.

A slick, readable, and practical 12-page two-color newsletter in the Peters style. Each issue has a theme with lots of short pieces on what different companies and people are doing to solve various business problems. Recommended.

Training
Lakewood Publications, Inc.
50 South Ninth Street
Minneapolis, MN 55402
(612) 333-0471, (800) 328-4329
Published monthly.
Subscriptions: United States one year $64, two years $108, and three years $138; Canada and Mexico one year $74; international one year $85.

Includes articles on issues of interest to managers and trainers, with frequent articles on quality management. Articles are practical, timely, well-written, and of interest to all involved in implementing TQM. Highly recommended.

Training & Development
American Society for Training and Development, Inc.
1640 King Street, Box 1443
Alexandria, VA 22313-2043
(703) 683-8100
Published monthly.
Subscriptions: Members of ASTD receive this magazine as part of their membership. Nonmember subscriptions available in the United States at $75 per year. Write for information on international subscription rates.

Includes articles on training, human resources, and management issues, with frequent coverage of quality management topics in these areas. Recommended.

Magazines and Journals on Quality in Health Care Management

Healthcare Forum Journal
The Healthcare Forum
830 Market Street
San Francisco, CA 94102
(415) 421-8810
Published bimonthly.
Subscriptions: Available as part of membership in The Healthcare Forum or one year $45, two years $80, three years $115. Canada/Mexico add $15 per year; one year all other countries $90.

A well-done journal that regularly includes articles on TQM. Articles are often in-depth yet down to earth and readable. Includes many ads on resources for health care managers looking to implement TQM. Recommended.

Healthcare Quality Abstracts
COR Healthcare Resource
P.O. Box 4095
Santa Barbara, CA 93140
(805) 564-2177
Fax (805) 564-2146
Published monthly except July.
Subscriptions: United States and Canada one year $98; all other countries one year $110.

This is just what it says it is—a newsletter format abstracting of current articles from a variety of journals all dealing with quality management in health care. The listing is broken up by topic and the sources for all articles are listed on the back page with addresses and phone numbers of publishers. Recommended (for those who can use it).

The Joint Commission Journal of Quality Improvement
Mosby-Year Book, Inc.
11830 Westline Industrial Drive
St. Louis, MO 63146-3318
(314) 453-4351, (800) 453-4351
Published monthly.

Subscriptions: $115 per year for United States, all other countries $125 per year.

Under the editorial direction of The Joint Commission, the goal of this refereed journal is to publish articles that emphasize the improvement of health care quality, which includes the measurement, assessment, and/or improvement of performance in health care quality and delivery. The journal includes how-to articles and case studies. Highly recommended.

Quality Management in Health Care
Aspen Publishers Inc.
7201 McKinney Circle
Frederick, MD 21701
(800) 638-8437
Published quarterly.
Subscriptions: $134 per year in the United States and Canada. For international subscriptions, contact: Swets Publishing Service, P.O. Box 825, 2160 SZ Lisse, The Netherlands.

This is a refereed journal with the purpose of providing a forum to explore the theoretical, technical, and strategic elements of quality management, and to assist those who wish to implement TQM in health care. Articles tend to be practical, somewhat technical, and oriented toward specific tasks in health care management. Recommended.

Magazines and Journals on Quality in Engineering

APICS-The Performance Advantage
American Production and Inventory Control Society, Inc.
500 West Annandale Road
Falls Church, VA 22046-4274
(800) 444-2742
Published quarterly.
Subscriptions: Included as part of membership package. Nonmembers subscriptions available at $30 per year; $40 in Mexico and Canada, and $50 for all other international subscriptions.

Covers the latest manufacturing principles and practices, case studies, columns, and news. Often includes articles on quality management in its field. Recommended.

Industrial Engineering
Institute of Industrial Engineers
25 Technology Park
Norcross, GA 30092
(404) 449-0461
Published monthly.
Subscriptions: Included as part of membership. Subscriptions available in the United States to nonmembers for one year $49, two years $82, and three years $115; Canada and international subscriptions one year $70, two years $120, three years $178.

While aimed at engineers, its articles are non-technical and oriented toward the application of TQM principles. Includes a regular monthly column on quality. Highly recommended.

Quality
Hitchcock Publishing Company
One Chilton Way
Radnor, PA 19089
Published monthly.
Subscriptions: United States one year $70. Canada and Mexico one year $85. International one year $160.

Includes articles that focus mainly on the technical aspects of implementing Total Quality Management in production and manufacturing environments. Includes reviews and event calendar. Highly recommended.

Quality and Reliability Engineering
John Wiley & Sons Ltd.
Baffins Lane, Chichester
Sussex PO19 1UD, England
Published bimonthly.
Subscriptions: one year $495.

A technical journal with articles designed to fill the gap between theoretical methods and scientific research on one hand and current industrial practices on the other. Higher specialized and mathematical. Recommended (only for corporate or university libraries for use by engineers in this area).

Total Quality Environmental Manager
John Wiley & Sons, Inc.
605 Third Avenue
New York, NY 10158
(212) 850-6479
Published quarterly.
Subscriptions: United States, Canada, Mexico one year $159; other countries one year $209.

While aimed at managers and, to some degree, engineers, in the environmental area, its articles are practical and cover a broad spectrum in the application of TQM principles to this field. Worth having in the library of any company interested in this area. Highly recommended.

Magazines and Journals on Quality for Other Professional Areas

Management Accounting
Institute of Management Accountants
10 Paragon Drive
Montvale, NJ 07645-1760
(201) 573-1760
Published monthly.
Subscriptions: Free to members; nonmembers $125 per year.

Frequently includes articles on activity-based costing and other subjects related to the implementation of quality principles in accounting processes. For accountants interested in TQM. Highly recommended.

Total Quality in Hospitality
Magna Publications Inc.
2718 Dryden Drive
Madison, WI 53704-3086
(608) 246-3580, (800) 433-0499
Published monthly.
Subscriptions: $198 per year, with additional subscriptions to the same organization available at a discount.

A monthly, eight-page, two-color newsletter covering specific ideas for implementing quality, case studies, and company profiles for the hospitality industry. Recommended.

TQM in Higher Education
Magna Publications Inc.
2718 Dryden Drive
Madison, WI 53704-3086
(608) 246-3580, (800) 433-0499
Published monthly.
Subscriptions: $129 per year with discounts available for additional subscriptions to the same location.

A monthly, eight-page, two-color newsletter covering ideas on applying TQM principles, case studies, information on TQM tools, and other practical articles on quality management in colleges and universities. Recommended.

APPENDIX 3

MAJOR QUALITY ASSOCIATIONS

This appendix provides an annotated listing of associations working on quality management and related issues such as ISO 9000 and training. It is organized by geographic area: the United States and Canada, Europe, and Asia and the Pacific.

United States and Canada

American Institute for
Total Productive Maintenance (AITPM)
P.O. Box 5097
Stamford, CT 06904 USA
(203) 846-3777, (800) 394-5772
Fax (203) 846-6883
This organization focuses on total productive maintenance (TPM) quality practices. It sponsors a conference in the fall and publishes the *TPM Newsletter*.

American Management Association (AMA)
135 West 50th Street
New York, NY 10020 USA
(212) 586-8100
Fax (212) 903-8168
The AMA has increasingly taken note of quality practices over the past several years, incorporating these new management methods into its seminars. It publishes *Management Review*, *Organizational Dynamics*, and *HR Focus*. It also sponsors seminars, and sells audio and video training programs.

American Production and Inventory Control Society (APICS)
500 West Annandale Road
Falls Church, VA 22046 USA
(703) 237-8344, (800) 444-2742
Fax (703) 237-1071
This has long been an important organization for manufacturing professionals, providing education and certification programs. In recent years its seminars, publications, and annual conferences have focused on just-in-time manufacturing practices, skills development in capacity management, materials requirements, production, and master production planning all through seminars, annual conferences, and local workshops around the United States. Among its publications is *APICS—The Performance Advantage*. APICS is organized through a network of local chapters.

American Productivity
and Quality Center (APQC)
123 North Post Oak Lane
Houston, TX 77024 USA
(713) 681-4020
Fax (713) 681-5321
APQC is a quality research center that accumulates information on best practices for all processes. It does research for clients on who has best practices, supplies copies of articles on processes, conducts seminars, and sells publications. It is the leading center for benchmarking activities in North America. It publishes *Continuous Journey* and sponsors an annual benchmarking award program.

American Society for Quality Control (ASQC)
P.O. Box 3005
611 East Wisconsin Avenue
Milwaukee, WI 53201-3005 USA
(414) 272-8575, (800) 248-1946
Fax (414) 272-1734
This is the most important quality association in the world, with the largest membership. It is also a co-administrator of the Malcolm Baldrige National Quality Award, and publisher of several major journals in the field, including *Quality Progress*, *Journal of Quality Technology*, *Technometrics*, *Quality Engineering*, and *The Quality*

Review. It sponsors an extensive series of seminars on quality-related issues (primarily focused on manufacturing but expanding to services), and is a major publisher of books on quality themes. It offers a widely respected certification program for quality assurance and other related quality topics, sells books by mail, has local chapters all over North America and a variety of technical divisions and committees. If you could only belong to one quality association, this is it.

American Society for Training and Development (ASTD)

P.O. Box 1443
1630 Duke Street
Alexandria, VA 22313 USA
(703) 683-8100
Fax (703) 683-8103

This is the major American association for those developing training programs. Training in quality management has been a concentration of its efforts. It hosts annual conferences in the spring and fall and publishes *Training & Development, Technical & Skills Training,* and *Info-Line.*

American Supplier Institute (ASI)

15041 Commerce Drive S., Suite 401
Dearborn, MI 48120-1238 USA
(313) 336-8877
Fax (313) 462-4500

This organization has skills in implementing Taguchi methods, quality function deployment (QFD), and Total Quality Management (TQM). It offers training programs, consulting, conferences, books and videos, and publishes *ASI Journal.*

ASQC Quality Management Division

611 East Wisconsin Avenue
Milwaukee, WI 53201 USA

This is an organization with over 30,000 members devoted to promoting quality management as a profession and body of practices and, through a group of committees, to do research on quality practices. It holds an annual conference in February and publishes the *Quality Management Forum* newsletter.

Association for Manufacturing Excellence, Inc. (AME)

380 West Palatine Road
Wheeling, IL 60090 USA
(312) 520-3282

Member companies use this organization to determine how to excell in manufacturing through the application of education, documentation, research, and sharing experiences. Publishes *Target* Magazine.

Association for Quality and Participation (AQP)

801-B West 8th Street, Suite 501
Cincinnati, OH 45203 USA
(513) 381-1959
Fax (513) 381-0070

This organization focuses on quality improvement through employee involvement, especially of self-managed teams, labor-management cooperation, redesign of work, and other studies on employee involvement. It sponsors conferences, maintains a library and research service on quality and employee issues, publishes the *Journal for Quality and Participation,* a newsletter, grants organizational team excellence awards, and sells resource materials. It is organized by local chapters in the United States and Canada. It also administers the Annual National Team Competition.

Associacion Quebecois de la Qualité

455 St. Antoine West, Suite 600
Montreal, Quebec H2Z 1J1, Canada
(514) 866-6696
Fax (514) 866-6724

This local quality association has over 2,000 members, holds an annual conference in June, and publishes *Forum Qualité* and *La Qualité Totale.*

Canadian Network for Total Quality

c/o Conference Board of Canada
255 Smyth Road
Ottawa, Ontario K1H 8M7, Canada
(613) 526-3280
Fax (613) 526-4857

This recently established organization promotes coordinated activities among members to foster the use of Total Quality Management practices.

Canadian Supplier Institute
Skyline Complex
644 Dixon Road
Rexdale, Ontario M9W 1J4, Canada
(416) 235-1777
This organization has skills in the application of Taguchi techniques, quality function deployment, Total Quality Management and other productivity tools in the area of manufacturing and production. It has training programs, does some consulting, offers an assortment of books and videos, and hosts conferences.

Conference Board, Inc.
P.O. Box 4026, Church Street Station
New York, NY 10261-4026 USA
(212) 759-0900
Fax (212) 980-7014
This nationally recognized council of American businesses is best known for its research on business trends. It also hosts an annual quality conference covering various tools and techniques in the general management of business. It publishes *Across the Board* magazine. It also publishes research papers, some of which are related to quality practices.

Federal Quality Institute (FQI)
441 F Street NW, Room 333
Washington, DC 20001 USA
(202) 376-3747 or 376-3753
Fax (202) 376-3765
FQI is the most important U.S. government organization involved in fostering the use of quality practices. It is responsible for helping all U.S. government agencies implement start-up activities in the area of Total Quality Management. It is a government-wide source for information on quality, using the FQI Information Network, which offers to agencies materials such as books, articles, videos, and case studies. It publishes Federal Quality News and administers the QIP Award and the Presidential Quality Award.

Fox Valley Quality/Productivity
Resource Center (Q/PRC)
Fox Valley Technical College
5 Systems Lane (Bordini Center)
P.O. Box 2277
Appleton, WI 54913-2277 USA
(414) 735-2277
Probably the best-known organization for quality practices and education among all community colleges in North America, this is a center for quality education, consulting, and research. Its library on quality management practices is extensive. Its mission is to help local business and nonprofit organizations improve their quality practices through speakers, training, and library resources.

Madison Area Quality
Improvement Network (MAQIN)
2909 Landmark Place, Suite 201
Madison, WI 53713 USA
(608) 277-7800
Fax (608) 277-7810
This very active quality organization is best known for hosting the annual Hunter Conference on Quality. Held usually in May, it runs several days, covers all aspects of quality implementation programs—both in the private and public sector—and is considered one of the more important national quality conferences. It also sponsors local seminars throughout the year.

Milwaukee First in Quality
Metropolitan Milwaukee Association
   of Commerce
756 North Milwaukee Street
Milwaukee, WI 53202 USA
(414) 287-4100
This is one of the largest and most active local quality organizations in the United States. It offers a variety of quality skill-building seminars and hosts a major annual convention in the area of quality. Particular focus has been in the area of implementing quality in manufacturing environments, although its offerings are expanding into the service sector.

Minnesota Council for Quality (MCQ)
2850 Metro Drive, Suite 300
Bloomington, MN 55425 USA
(612) 851-3181
Fax (612) 851-3183
This is a very active state council that promotes quality in Minnesota through conferences, seminars, other educational programs, administers the Minnesota Quality Award and the Quality Service Award; it is a model regional quality council.

National Institute of Standards and Technology
Route 270 and Quince Orchard Road
Administration Building, Room A537
Gaithersburg, MD 20899 USA
(301) 975-2036
Fax (301) 948-3716
This organization, along with the ASQC, manages the Malcolm Baldrige National Quality Award. It is responsible for defining the criteria and point values for the award and also for the judging process.

**National Society for Performance
and Instruction (NSPI)**
1300 L Street NW, Suite 1250
Washington, DC 20005 USA
(202) 408-7969
Fax (202) 408-7972
This is a large organization, with members from many countries, dedicated to improving productivity and effectiveness in the workplace. Over 600 chapters meet monthly. Oone of the more important topics is the science of performance technology, an organized process for solving problems concerning performance. It holds an annual conference in April and publishes *Performance & Instruction* and *Performance Improvement Quarterly*.

Philadelphia Area Council for Excellence (PACE)
1234 Market Street, Suite 1800
Philadelphia, PA 19107-3718 USA
(215) 972-3977
Fax (215) 972-3900

PACE promotes the application of Total Quality Management and Deming approaches. It is one of the oldest regional quality organizations.

Quality & Productivity Management Association (QPMA)
300 North Martingale Road, No. 230
Schaumburg, IL 60173 USA
(708) 619-2909
Fax (708) 619-3383
This network of business professionals focuses on continuous improvement strategies that can be applied in the workplace. It conducts educational programs, workshops, distributes publications, and administers the Annual QPMA Leadership Award.

Standards Council of Canada
45 O'Connor Street, Suite 1200
Ottawa, Ontario K1P 6N7, Canada
(613) 238-3222
Fax (613) 995-4564
This organization acredits quality certification associations and companies. It coordinates the activities of Canada's ISO Technical Committee responsible for ISO certification activities.

Europe

Asociacion Española Para la Calidad (AECC)
92 Calle Zurbano, 1-D
28003 Madrid, Spain
34/1 441 7777
Fax 34/1 441 777733
This is a national quality organization that promotes quality practices, offers educational programs, and testing, and maintains ties to quality associations in Latin America.

**Association Française
de Normalisation (AFNOR)**
Tour Europe, Cedex 7
92049 Paris La Defense
France
33 1 42 91 55 55
Fax 33 1 91 56 56

AFNOR is the national ISO member organization representing France. It is responsible for promoting the use of ISO standards by French companies.

British Deming Association
2 Castle Street
Salisbury, Wiltshire SP1 1BB England
44 7 22-412-138
Fax 44 7 22-331-313
Its purpose is to promote Dr. Deming's view of quality practices. It has both individual and corporate members and holds an annual conference in April.

British Quality Association (BQA)
P.O. Box 712
61 Southwark Street
London SE1 1SB, England
44 71 401-2844
Fax 44 71 401-2715
This is a large organization and is the administrator of the British Quality Award. BQA promotes TQM principles, consults, holds seminars, training programs and seminars. It is one of the major European quality organizations.

British Standards Institution (BSI)
2 Park Street
London W1A 2BS England
44 7 1 629-9000
Fax 44 7 1 629-0506
The primary purpose of this organization is to administer the ISO 9000 certification process in the United Kingdom.

Deutsche Gesellschaft fur Qualitat
August-Schanz Strasse 21A
P.O. Box 50 07 63
6000 Frankfurt am Main 50
Germany
49 6 995 4240
Fax 49 6 995 424133
This is the major German society for quality, has both corporate and individual members, and provides quality programs nationally.

Deutsches Institut fur Normung (DIN)
Burggrafenstrasse 6
Postfach 1107
1000 Berlin 30, Germany
49 3 026 011
Fax 49 3 026 01231
This is Germany's national ISO member institution, responsible for the implementation of ISO 9000 certification programs in the country.

European Foundation for Quality Management (EFQM)
Avenue des Pleiades 19
B-1200 Brussels, Belgium
32 3 775-3511
Fax 32 2 779-1237
The EFQM supports the use of quality practices by Western European companies. Membership is by company. It manages the European Quality Award program, begun in 1992; hosts conferences and education programs; and publishes *Quality Link* Newsletter.

International Academy for Quality (IAQ)
P.O. Box 50 07 63
50 Frankfurt am Main W6
Germany
49 6 954 80010
Fax 49 6 954 800133
To be a member one must be sponsored by three or more scholars operating on at least two continents. Members encourage international understanding of quality issues, promoting research on the philosophy and practice of quality. It also administers the IAQ Award.

Italian Association for Quality (AICQ)
Piazza A. Diaz 2
20121 Milan, Italy
39 2 720 03460
Fax 39 2 720 23085
The Associazione per le Qualita promotes the use of quality practices in Italy; its members are corporate enterprises, and it publishes *Qualita*.

Mouvement Français Pour la Qualité (MFQ)
5, esplanade Charles de Gaulle
92733 Nanterre cedex, France
33 1 40 97 06 40
Fax 33 1 47 25 32 21
This is an important French quality association with the mission of promoting exchange and application of information on quality practices by French organizations and companies. It offers training, conferences, runs regional quality circles, administers the Prix Nationale de la Qualité and publishes *Qualité en Moevement* and *Regard sur la Qualité.*

National Quality Information Centre
P.O. Box 712
61 Southwark Street
London SE1 1SB, England
44 71 401-7227
Fax 44 71 401-2725
This very large British quality organization promotes the use of quality assurance practices in British industry through education programs, consulting, publications, and research done jointly with universities. It publishes *Quality News* and the *Quality Forum.*

Asia and the Pacific

Asian Productivity Organization (APO)
8-4-14 Akasaka, Minato-ku
Tokyo 107, Japan
81 3 3408 7221
Fax 81 3 3408 7220
This is an intergovernment organization with representatives from 18 governments from Asia and the Pacific. It surveys on best practices, offers training programs, and publishes on quality topics. Its purpose is to accelerate economic development through improved economic productivity; this is a major international quality information clearinghouse.

Australian Institute of Management (AIM)
215 Pacific Highway
North Sydney NSW 2060
Australia
61 2 929 7922
Fax 61/2 922 2210
While a traditional management development association with a large membership, it increasingly has been exposing its membership to quality-oriented management practices. It publishes *AIM News* and *Management Review.*

Australian Organization for Quality
27 Palmerston Crescent
South Melbourne VIC 3205
Australia
61 3 699 4144
Fax 61/3 696 4510
This is the oldest and most-recognized quality organization in Australia. It promotes quality practices through education programs and support services.

Australian Quality Council
P.O. Box 298
St. Leonards NSW 2065 Australia
02 439-8200
Fax 02-906-3286
This is an umbrella organization combining the Enterprise Australia, Ltd., Australian Quality Awards Foundation, Total Quality Management Institute, and the Quality Society of Australia. It hosts seminars and courses on quality, does evaluation consulting, and publishes various resources including *The Quality Magazine* and the guidelines for the Australian Quality Awards.

Japan Management Association (JMA)
Nihon-Nohritsu Kyokai Bldg.
3-1-22 Shiba-koen, Minato-ku
Tokyo 105, Japan
81 3 3434 6211
Fax 81 3 3434 1087
JMA offers courses, seminars, and conferences on all aspects of business; conducts trade shows and international visits; publishes extensively on all manner of management topics—*JMA Management News, JMA Journal,* and *Human Resource Development*—and administers awards of excellence. Membership consists of over 2,500 corporations.

Japanese Standards Association (JSA)
1-24 Akasaka 4-chome
Minato-ku
Tokyo 107, Japan
81 3 3583 8008
Fax 81 3 3583 0698
JSA promotes standardization on quality issues and offers education, all as an organization under the control of the Ministry of International Trade & Industry. It also manages many of the ISO functions in Japan, sponsors a national conference in May, publishes *Standardization & Quality Control* and *Standardization Journal*, and administers the JSA Excellent Standardization Document Award.

Union of Japanese Scientists and Engineers (JUSE)
5-10-11, Sendagaya, Shibuya-ku
Tokyo, 151 Japan
81 3 5379-1227 or 1231
Fax 81 3 3225-1813
This organization is best known as the administrators of the Deming Prize, Japan's highest quality award. It is Japan's leading quality organization and provides training, hosts symposia, consults, maintains a library, and publishes a variety of magazines and journals, including the monthly *Total Quality Control* and *Engineers*, and, every two months, *Societas Qualitatis* that describes activities of JUSE.

Korean Society for Quality Control
5-2 Soonhwa-dong, Chung-ku
Seoul, South Korea
This organization conducts research on quality practices, offers educational programs, publishes and distributes materials on quality, and provides channels of communication between local businesses and academia.

Korean Standards Association (KSA)
5-2 Soonhwa-dong, Chung-ku
Seoul 100-130, South Korea
82 2 772 3417
Fax 82 2 754 0346
This is the national industrial standardization organization for South Korea. It offers a full compliment of courses and seminars, along with consulting, to organizations seeking quality certifications. It hosts a national annual conference held in either October or November, and publishes *Quality and Management*, *Standardization*, and *Factory Management*.

New Zealand Organization for Quality (NZOQ)
P.O. Box 622
Palmerston North
New Zealand
64 6 3569 0099
Fax 64 6 350 5604
This is the major quality organization for New Zealand, promoting quality practices as a means of improving New Zealand's competitive posture in the world economy. It hosts a national conference in May, and publishes *Quality New Zealand*, *Q-NewZ*, and *Software Q-NewZ*.

Singapore Quality Institute (SQI)
Blk. 18, Ngee Ann Polytechnic
No. 03-08 Clementi Road
2159 Singapore
65 467 4225
Fax 65/467 4226
Like the American ASQC, it promotes quality practices, particularly in business, through seminars, conferences, and publications; it is a clearinghouse for local information on quality. It publishes *QC Focus* and the *SQI Yearbook*.

INDEX